AN EARLY MIDDLE EOCENE FLORA FROM THE YELLOWSTONE-ABSAROKA VOLCANIC PROVINCE, NORTHWESTERN WIND RIVER BASIN, WYOMING

BY

H. D. MacGINITIE

with chapters by Estella B. Leopold and W. L. Rohrer

UNIVERSITY OF CALIFORNIA PUBLICATIONS IN GEOLOGICAL SCIENCES

Volume 108

UNIVERSITY OF CALIFORNIA PRESS

AN EARLY MIDDLE EOCENE FLORA FROM THE YELLOWSTONE-ABSAROKA VOLCANIC PROVINCE, NORTHWESTERN WIND RIVER BASIN, WYOMING

AN EARLY MIDDLE EOCENE FLORA
FROM THE YELLOWSTONE-ABSAROKA
VOLCANIC PROVINCE, NORTHWESTERN
WIND RIVER BASIN, WYOMING

BY

H. D. MacGINITIE

with chapters by Estella B. Leopold and W. L. Rohrer

UNIVERSITY OF CALIFORNIA PRESS

BERKELEY · LOS ANGELES · LONDON

1974

University of California Publications in Geological Sciences

Volume 108

Approved for publication February 16, 1973

Issued April 1, 1974

University of California Press
Berkeley and Los Angeles
California

❖

University of California Press, Ltd.
London, England

CONTENTS

[v]

Contents

AN EARLY MIDDLE EOCENE FLORA FROM THE YELLOWSTONE-ABSAROKA VOLCANIC PROVINCE, NORTHWESTERN WIND RIVER BASIN, WYOMING

BY

H. D. MacGINITIE, with chapters by ESTELLA B. LEOPOLD and W. L. ROHRER
(A contribution of the Museum of Paleontology, University of California, Berkeley)

ABSTRACT

THE FOSSIL FLORA described occurs near the base of the Middle Eocene volcaniclastic deposits along the north border of the Wind River Basin in northwestern Wyoming. The volcanic deposits are part of the great Absaroka-Yellowstone volcanic field. The flora was, in part, deposited in the floodplain of a moderately large river which came into the area from the northwest. In addition to abundant megafossils the floral zone also yields a large and well-preserved pollen assemblage. The present vegetation of the area ranges from a taiga-type forest along the south border of the Absarokas to semi-arid grass and sage brush typical of the middle Rocky Mountain basins. The climate is of the extreme continental type with scanty rainfall, cold winters, and large annual temperature ranges.

Fifty-four fossil species are listed, of which 20 are new. In addition to 5 ferns, 1 horsetail, 2 gymnosperms, 3 palms, 2 herbs, and 5 species of doubtful affinity, the fossil flora contains 36 well-defined, woody dicots. Fifty-five percent of the dicot leaves were non-entire; and 61 percent, in terms of the related living species, were deciduous. A large proportion of the living species related to the fossil species are inhabitants of low ground (floodplain) areas. Only a few species can be related to living plants of upland or hillside areas. The 11 most abundant genera (10 or more specimens collected) in approximate order of abundance are: *Platanus, Lygodium, Canavalia, Populus, Cedrela, Sabalites, Symplocos, Ilex, Acalypha, Dendropanax* and *Aleurites*. The flora has a distinct tropical American aspect. Among the genera indicative of tropical or near-tropical conditions are: *Acalypha, Acrostichum, Apeiba, Canavalia, Cedrela, Eugenia, Dendropanax, Machilus-Persea, Luehea, Saurauia, Schefflera* and *Sterculia*.

The volcaniclastic series has furnished K/Ar dates at several horizons. These dates show that the age of the floral horizon (Kisinger Lakes-Tipperary) is from 48 to 49 my. The flora forms one member of a sequence in the Middle Rocky Mountains: Wind River, 52 my.; Kisinger Lakes, 48.5 my.; Green River, 46 my.; and Florissant, 36 my. There is a trend from humid, warm temperate to dryer and more tropical conditions, from the Early Eocene to the Middle Eocene. The floras furnish critical evidence concerning the Paleogene development of vegetation in the middle Rockies.

Although there exists no living floristic group closely similar to that of the fossil flora, the present semi-deciduous flora of southwestern Mexico, the bosque tropical subdeciduo, has many resemblances both floristically and vegetationally. The present vegetation and climate of the area in the vicinity of Tepic, in western Mexico, corresponds reasonably well with that of the early Middle Eocene in the Wind River Basin. The climate indicated by the fossil plants was tropical or near-tropical, with a pronounced winter dry season. The annual precipitation was between 35 and 55 inches. The average annual temperature was between 19°C and 23°C. The average temperature of the coldest month was not below 15°C. The climate was either frostless, or, possibly, with rare, light frosts.

The fossil flora shows how extraordinary have been the climatic changes in the area since the Middle Eocene. The average annual temperature has decreased by approximately 20°F; absolute minima have declined by 50°F to 60°F; annual rainfall has decreased by 20 to 40 inches. The elevation has increased by 4000 to 8000 feet or more.

INTRODUCTION

THE FOSSIL FLORA treated in this paper is found at two localities along the northwestern margin of the Wind River basin, northwestern Wyoming. Large collections made during the years 1965–1969 are stored in the Museum of Paleontology, University of California, Berkeley. The two sites are separated by an east-west distance of approximately 30 miles. The first locality discovered was that near Tipperary, an old post office, now abandoned, about 20 miles north-northwest of Crowheart, Wyoming. The fossil plants occur in a tuff lens in the SW ¼ of Section 18, T 6 N, R 4 W, Wind River Meridian, Fremont County. Crowheart is on highway 26-267, 29 miles southeast of Dubois. The first collection from Tipperary was made by Mr. N. H. Brown of Lander, Wyoming. This was sent to E. W. Berry of the United States Geological Survey who described the flora in his paper, "A flora of Green River age in the Wind River basin of Wyoming" (1930*b*). He listed 42 species (p. 60), 9 of which were small seeds, calyces, and similar objects. Of the remaining 33 fossil species about 20 were either wrongly identified or named as form genera according to the paleobotanical philosophy of the time: *Ficus, Aralia, Laurus* and similar epithets. The revision of Berry's list may be found below on page 20. He overemphasized the relation of the flora to that of the Green River beds in northwestern Colorado, but, in spite of the inadequate identifications, his reconstruction of the paleoecology (pp. 58–60) seems to approach rather near the truth as I see it, although the flora is more tropical than he estimated.

The original Kisinger Lakes locality was discovered by W. L. Rohrer of the United States Geological Survey during his geologic mapping of the Kisinger Lakes quadrangle in the years 1961–1964. The stratum containing the flora crops out approximately along the 9200 foot contour in the south half of Section 11, T 43 N, R 109 W. The site can be reached by a difficult jeep trail which branches north from the forest logging road in Section 24.

Two additional localities in the Kisinger Lakes floral horizon were investigated in the summer of 1971. The first is in the approximate center of Section 11, T 43 N, R 109 W in a slumped area from the main horizon just above. The second is about 6 miles east of the original locality in the Esmond Park quadrangle at the headwaters of the East Fork of Sixmile Creek and at the south foot of Ramshorn Peak, SE ¼, Section 11, T 43 N, R 108 W. Fossil plants are abundant at each locality.

ACKNOWLEDGMENTS

A paper of this kind must necessarily be a cooperative effort. I owe heartfelt thanks to many people some of whom I shall not name. I am grateful to Mr. W. L. Rohrer, Dr. Richard Keefer and Dr. David Love for help with stratigraphic problems; to Dr. John Obradovich for information on K/Ar dates. Mr. Howard Schorn, Dr. Jack Wolfe and Dr. Leo Hickey spent much time in assisting with taxonomic problems. Dr. Estella Leopold furnished the chapter on the palynology. Mr. W. L. Rohrer wrote the chapter on the stratigraphy of the area, helped with collecting in the field on many occasions and with the solution of stratigraphic problems. I owe especial thanks to the National Science Foundation which has financed the major part of the work.

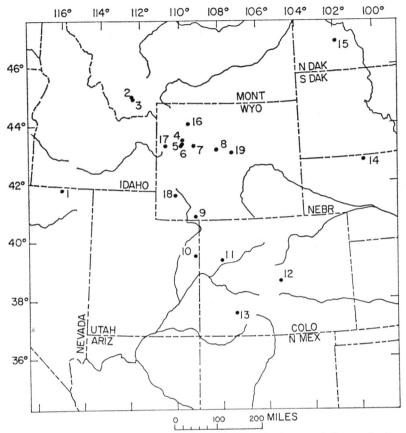

Fig. 1. Index map showing localities of floras mentioned in the text. 1, Copper Basin; 2, Ruby; 3, Mormon Creek; 4, Kisinger Lakes; 5, Wind River; 6, Coyote Creek; 7, Tipperary; 8, Boysen; 9, Little Mountain; 10, Rainbow; 11, Wardell Ranch; 12, Florissant; 13, Creede; 14, Kilgore; 15, Golden Valley; 16, Valley; 17, Rhodes Ranch; 18, Fort Hill; 19, Schoening.

GEOLOGIC RELATIONS

The Laramide Revolution of the latest Cretaceous, Paleocene, and Early Eocene (Keefer, 1970:10; Love 1970:115) drained the great Cretaceous inland sea and brought into being the complex Rocky Mountain orogenic belt. This is characterized by extended mountain ridges and intervening basins roughly aligned in a northwesterly direction in northern Wyoming. The ranges tend to be asymmetrical with a relatively gentle slope on one side and a much steeper slope on the other, which is usually marked by overturning and thrusting (?). The western Wind River Basin is bordered on the north by the steep slopes of the Owl Creek Mountains and on the southwest by the gentle, extended, northern slopes of the Wind River range. That is, the steep and partially overturned sides of both the Owl Creek and Wind River ranges face toward the southwest. The basin is typical of the great, intermontane basins in Colorado and Wyoming (see fig. 2).

In describing the Wind River Basin, Keefer writes (1965*b*:1878): "This extensive structural depression is completely surrounded by broad belts of folded and faulted Paleozoic and Mesozoic rocks: the Granite Mountains on the south;

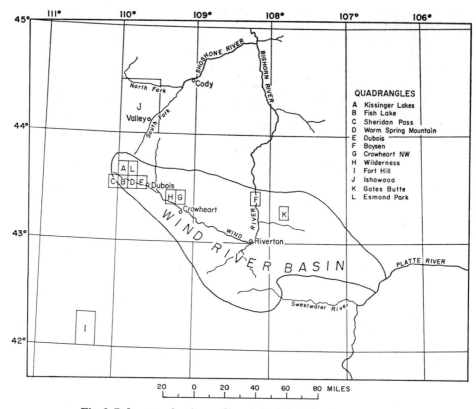

Fig. 2. Index map showing outline of the Wind River Basin and the
location of quadrangles mentioned in the text.

the Wind River Mountains on the west; the Washakie, Owl Creek and southern
Big Horn Mountains on the north; the Casper Arch on the east; and the Laramie
Mountains on the southeast. The central part of the basin is covered by nearly
flat-lying lower Eocene rocks that overlap all older strata near the basin margins.
The structural basin includes about 8,500 square miles." The Laramide crustal
movements consisted not only in upthrusting of the ridges but in the sinking of
the intervening basins (Keefer, 1965a:58–60). The maximum stratigraphic difference between the bordering uplifts and the basin floor in the Wind River
structure is about 35,000 feet (Keefer, 1965b:1889) with approximately half
of this difference due to sinking of the valley floor relative to sea level. This
process is reflected in many other basins of the Rocky Mountain province. Thus
the basin became a broad depositional axis, receiving sediments from the bordering uplifts, from nearly the beginning of the Tertiary until some time in the
Middle or Late Pliocene when uplift initiated the present degradation cycle. The
outline of the basin is an elongated oval, approximately 190 miles in length and
75 miles in maximum width, and trending northwesterly. During the time of
maximum orogenic change, mainly in the Early Eocene, or early Wasatchian
time, deposition was sporadic but became fairly continuous after the Gray Bullian
and continued into the latest Eocene. The basin sediments contain abundant

plant and animal fossils in many places. The northeastern part of the basin, north and northeast of Lysite, has been a mine of Paleogene mammals and is the type locality for the Lysite and Lost Cabin mammal horizons. The Rate Ranch megaflora of probable middle Eocene age occurs in the same general area. In the western part of the basin are the localities for several Eocene floras: Schoening, Tipperary, East Fork, Boysen, Kisinger Lakes, Wind River, Coyote Creek, Three Tarns and others. The approximate locations of the fossil plant cites are given on the map (fig. 1), and their ages in the correlation chart (Table 4).

The Eocene sediments of the Wind River Basin are divided into 5 formations in ascending order: Indian Meadows, Wind River, Aycross, Tepee Trail and Wiggins.

The Indian Meadows Formation marks the initial stage of basin filling. It "is composed of variegated brick-red, purple, gray and white claystone, siltstone, sandstone and conglomerate, and thin beds of gray algal-ball limestone" (Keefer, 1965a:39). The conglomerate beds are composed for the most part of rounded pebbles and cobbles of resistant Precambrian and Paleozoic rocks. These sediments were deposited during and just after the Early Eocene orogeny, mostly as fans originating in the bordering ranges. The Indian Meadows Formation is coarse and conglomeratic near the uplifts but is composed of finer material toward the depositional center of the basin. The most widespread and characteristic portion of the Tertiary sedimentary sequence is the Wind River Formation, mostly of later early Eocene age (Lost Cabinian) (Keefer, 1957:188; 1965a:44; 1969:25). This crops out over large areas in the basin and varies from about 1000 to over 7000 feet in thickness; "it is the surface rock in nearly 4,000 square miles of the central part of the basin" (Keefer, 1965a:44). In the northwestern portion of the basin the lower part of the Wind River Formation recalls the typical characteristics of the Wasatch Formation in southern Wyoming—red-banded, often with brilliant coloring. The lower 400 to 600 feet is comprised of "red siltstone and shale, soft to hard buff fine-grained shaly sandstone, and white to buff cobble and pebble conglomerate" (Keefer, 1957:190). The lithology of the formation varies from place to place. "In general, however, two facies predominate—a coarse boulder facies representing deposition along the mountain slopes and a fine-grained, commonly brightly variecolored facies representing deposition farther out in the basin" (Keefer, 1965a:44–45). The lithology of the Wind River Formation along the slopes of the range south of the Wind River differs considerably from that of the formation on the northwestern side of the basin, which has considerably more volcanic debris and is finer grained. There is a sequence of tuffaceous drab beds toward the middle of the formation northwest of Dubois, and it is in this portion of the formation, in fine-grained, tuffaceous layers, that the remarkably rich and well-preserved Wind River flora is found. Striking exposures of the lower part of the formation are to be seen on the north side of the highway both northwest and southeast of Dubois, where the highly colored, red-banded beds crop out, eroded into badland topography. The change from drab to reddish sediments occurs as the beds approach the brilliantly colored Chugwater Formation of Triassic age which crops out on the lower slopes of the Wind River Range southwest of Dubois.

Above the Wind River Formation three later formations were recognized by Love along the southern margin of the Absaroka Range, the western end of the Owl Creek Mountains and the northwestern part of the Wind River Basin (Love, 1939:11, 66–86). These are, from older to younger the: Aycross, Tepee Trail and Wiggins Formations. The Wind River Formation contains tuffaceous strata, especially toward the top, but the succeeding deposits show increasing proportions of volcaniclastic materials until the great pile of tuffs and agglomerates comprising the Wiggins Formation is reached. Thus all five formations are typical basin deposits modified more and more by volcanics toward the top of the sequence.

The Aycross Formation was the name proposed by Love (1939:66) "for the middle Eocene sequence lying unconformably upon all older rocks from pre-Cambrian to lower Eocene in age, and which is unconformably overlain by strata of Upper Eocene age." He identified Aycross strata along the southern margin of the ranges from Bear Creek to Tipperary. The sediments assigned to this group are complex and variable, changing in character both horizontally and vertically. Love states (1939:66–67) "the physical appearance of the Aycross formation changes so rapidly in a short distance that even a general description at the type area on North Mesa does not apply 4 miles to the south on Coulee Mesa." The frequent presence of conglomerate lenses, especially in the lower part of the Aycross, indicates that the major part of the sediments are deposits of streams originating in the bordering uplands on the north and northwest. The formation at the type locality is tuffaceous throughout, and most of the volcanic debris is water-laid. The higher ground to the north was undergoing up-building by intense volcanism; the increasing energy of the streams was probably due in great part to this factor, although sporadic uplifts may have occurred. At North Mesa Love made three divisions of the Aycross: (1) at the base, about 200 feet of "soft, brilliantly variegated clays in which red, purple and green dominate. A considerable amount of fine andesitic volcanic tuff is present" (1939:67); (2) a conspicuous lens of greenish tuff, about 50 feet thick, containing rounded cobbles of red, andesitic porphyry from 2 to 12 inches in diameter; (3) about 750 feet of "variegated clays, shales, sandstones, conglomerates and volcanic rocks." The fossil flora described by Berry comes from near the base of the Aycross about 9 miles southeast of the type locality on North Mesa. This is the Tipperary site in Section 18, T 6 N, R 4 W. Above the Aycross is the Tepee Trail Formation, defined by Love (1939:73) as "the Upper Eocene sequence unconformably overlying the Aycross Formation and all older rocks, and, in turn, unconformably overlain by Oligocene ? rocks." It "has a wider areal distribution than any other Cenozoic formation in this region." The formation is well bedded for the most part, green or brown in color, and "is made up almost entirely of basic andesitic material. To the north and northwest the formation is characterized by conglomerates and coarse breccias which gradually change southeastward to fine-grained, waterlaid tuffs." In the Kisinger Lakes Quadrangle the Tepee Trail Formation, as mapped by Rohrer (1966) "is about 1,800 feet thick and consists mainly of grayish-green fine-grained fluvial volcanic sedimentary rocks. The strata are interbedded claystone, siltstone, sandstone, bentonite and mudstone." A volcanic conglomerate about 18 feet thick, containing "green, gray and pink biotitic lapilli and chert and quartzite pebbles in a green tuffaceous sandstone matrix" is taken as the base of the formation. "This basal unit is ledge forming and disconformably (?) overlies a greenish-gray volcanic sandstone . . . assigned to the Wind River Formation" (Rohrer and Obradovich, 1969:58). This contact appears to be mainly lithological since there is no good evidence for any marked disconformity at this level in the area. The upper portion of the Wind River Formation just south of the Kisinger Lakes Quadrangle commonly contains much volcanic debris and it appears to grade upward into the "Tepee Trail Formation." "The upper 600 feet [of the Tepee Trail] contains may tongues of volcanic mudflow conglomerate and mudstone interbedded with the claystones and sandstones" (1969:58). The coarser materials grade basin-ward into siltstone and claystone.

Vertebrate fossils discovered by members of the departments of paleontology of the University of California, Berkeley, and the American Museum of Natural History, during recent investigations of the type locality of the Tepee Trail Formation, are Bridger E or Uinta A. These two faunal zones may overlap and probably need more clarification. Former usage might designate this horizon as late Middle Eocene.

Following the Tepee Trail Formation is the Wiggins Formation whose resistant strata form the imposing elevations of the Absaroka Mountains in the area just southeast of Yellowstone Park. The Pinnacle Buttes, the Ramshorn and other elevations are a characteristic feature of the landscape in this area. "The Wiggins Formation at Pinnacle Buttes is an alternating sequence of brownish-gray volcanic mudflow conglomerate (about 85 percent), bluish-gray rounded volcanic stream conglomerate (about 10 percent) and fluvial tuff (about 6 percent). The mudflow conglomerate tends to be massive and in beds, commonly more than 20 feet thick, that consist of varicolored angular andesite and basalt as much as 8 feet in diameter in a tuff matrix. . . . Distinctive white bands, generally less than 6 feet thick, in the dissected Absaroka escarpment (fig. 3) are volcanic ash." (Rohrer and Obradovich, 1969:59–60). There is apparently an unconformity at the base of the Wiggins since the Tepee Trail and Wiggins intertongue and the contact migrates stratigraphically upward toward the southeast (1969:61). The Wiggins Formation varies in thickness from 1500 to 4000 feet.

The Tipperary locality is on the north side of the Wind River Basin, on the south border of the western extension of the Owl Creek Mountains. The sediments are typical basin deposits varying in character both vertically and horizontally, and marked by channeling. The layer containing the fossil plants is a lens of fine-grained, sandy tuff, varying from a tuffaceous siltstone to a tuffaceous sandstone, whose present outcrop covers only a few acres. This tuffaceous lens is from 75 to 100 feet above the base of the Middle Eocene Aycross Formation (Love, 1939: 69). At this locality the Aycross beds are deposited on a rather irregular surface of the Wind River Formation, which is composed of nearly flat-lying, poorly consolidated shales, silts and tuffaceous sandstones, which weather to badland topography. Above the unconformity are from 40 to 65 feet of dark, sandy and pebbly, tuffaceous beds, overlain by from 3 to 4 feet of leaf-bearing, resistant tuffs. Above the tuff lens is about 15 feet of weakly consolidated sands, followed by another thin tuff bed containing a few plant fossils. Above this, forming the surface layer, is a few feet of soft, thin-bedded sandstone. The lower tuff lens is full of leaf fossils, especially palms, *Platanus, Cedrela,* and *Lygodium.* The leaves are not aligned with the bedding but show evidence of transport and are often curled and folded. It is difficult to obtain whole or even approximately complete leaves.

The Kisinger Lakes flora is found in a zone from 260 to 270 feet above the White Pass bentonite bed (see p. 14) in the Kisinger Lakes quadrangle. There are also a few poorly preserved leaves at 800 and 875 feet above the base of the "Tepee Trail." the fossil flora is preserved in a hard, tuffaceous siltstone layer from 2 to 3 feet thick. The zone containing the fossil leaves is approximately along the 9200-foot contour in the south half of Section 11, T 43 N, R 109 W. The flora is a typical floodplain flora, the megafossils largely comprised of trees whose living relatives are characteristic of floodplain habitats. Species which might be assigned to upland situations are rare. The pollen flora contains grains of fir, spruce, pine and oak, which may have grown on the moderate elevations to the northwest.

Since the stratum bearing the Kisinger Lakes flora is contained in generally fine-grained fluvial deposits, tuffaceous standstones, fine-grained tuffs, and a few interbeds of pebble conglomerate, this indicates that these river deposits were laid down in the floodplain of a moderately large stream of gentle gradient which came from the northwest or west. The floodplain must have been of considerable width, probably several miles, judging from the sediments. Love, in his paper on the Granite Mountains to the southeast (1970:118–119) says, "Middle Eocene was another time of crustal stability in most of central Wyoming. The rate of deposition around the Granite Mountains slowed to a minimum for Tertiary time in central Wyoming. Fundamental changes in drainage took place. For the first time the Wind River basin was filled with sediment, and streams from the Absaroka volcanic area flowed southeast along the west margin of the Granite Mountains and out into the Great Divide basin. . . . The Absaroka centers 90 miles away furnished a large amount of airborne and waterborne volcanic debris that is conspicuous throughout the western part of the Granite Mountains area." This "Kisinger River" may have persisted into the Late Eocene. The headwaters of the stream (or streams) may have been in the area of the Yellowstone volcanic centers to the northwest or more probably farther west in the Great Basin rise, perhaps in the area of the present Idaho Batholith.

The relations among the various Eocene deposits along the northwest margin of the basin are complex. After Gray Bull time the great volcanic field of the Yellowstone-Absaroka area became increasingly active. This became the source area of relatively enormous amounts of volcanic debris which was spread by wind and streams into the basin. Streams flowing out of the volcanic area increased their gradients and deposited interfingering, deltaic, fan, and floodplain sediments. There exist marked and alternating changes in lithology and facies from northwest to southeast. The various stream deposits are not necessarily of exactly the same age in the same general stratigraphic interval. Vertical and horizontal changes in lithology occur within a few miles. All the deposits of the sedimentary interval between the Wind River and Wiggins Formations were designated as Tepee Trail by Rohrer on the Kisinger Lakes Quadrangle geologic map (1966).

Recently some critical K/Ar ages have been determined from the basal Wiggins in the Pinnacle Buttes area (Rohrer and Obradovich, 1969:61). These are based on hornblendes from the andesitic volcanics and give ages of 46.2 ± 1.8 m.y. and 46.5 ± 2.3 m.y. which places at least part of the Wiggins as late Middle Eocene. Also dates were obtained from volcanic material less than 100 feet below the fossil plant horizon in the Kisinger Lakes Quadrangle and these, in round number were 48 to 49 m.y. Thus the "Tepee Trail" Formation as mapped in the Kisinger Lakes Quadrangle is entirely Middle Eocene and the floral horizon is early Bridger. It is probable that the lowest part of the "Tepee Trail" extends down into the Early Eocene.

From the results of K/Ar dating given above it appears that at least part of the stratigraphic interval in the Kisinger Lakes Quadrangle between the Wind River Formation and the Wiggins Formation may correspond to what Love called the Aycross Formation farther east along the margin of the Absarokas. Love's investigations along the southern margin of the Absarokas did not extend as far west as the Du Noir area and geologic mapping has not yet bridged the area between outcrops mapped as Aycross and the area of the Kisinger Lakes Quadrangle. Therefore, to avoid ambiguity, Rohrer in his chapter on the stratigraphy, has chosen to divide the deposits above the Wind River and below the Wiggins into an upper and lower volcaniclastic series.

Since there seems to be, at present, no general agreement as to the placement of the contact between the Wind River Formation and the overlying beds in the area, the writer suggests the use of the White Pass bentonite bed as the marker for the contact. This bed of altered tuff is widespread in the area and is easily recognized. Rohrer's stratigraphic diagram (fig. 3 p. 11 below) shows his interpretation of the lower volcaniclastic unit and the beds containing the Wind River flora.

[Mr. Rohrer has kindly furnished the following account of the stratigraphic relations among the fossil plant horizons in the Paleogene of the western Wind River basin, based on over eight years of geologic research and mapping in the area. Publication of the material is authorized by the Director, United States Geologic Survey. H.D.M.]

STRATIGRAPHY AND STRATIGRAPHIC RELATIONS
OF THE FOSSIL FLORAS
by W. L. ROHRER

GEOLOGIC SETTING AND CORRELATIONS

RECENT MAPPING in the Sheridan Pass, Fish Lake and Kisinger Lakes quadrangles has disclosed the presence of early and middle Eocene floras in fluvial volcaniclastic rocks in the eastern Gros Ventre and northwestern Wind River Basins. The lithology of these rocks indicates three source areas: the Wind River Range, the Jackson Hole area, and the Yellowstone-Absaroka volcanic province. The nearest source, the Wind River Range, was stripped of its Mesozoic and much of its Paleozoic rocks during the Late Cretaceous, post-Mesaverde time, and uplift of the range near the end of Paleocene time resulted in deposition of arkosic sediments in adjacent areas. Deposition of arkosic sediments continued through Early Eocene time. A series of uplifts in the Jackson Hole region, the second source area, during late early Eocene time resulted in deposition of quartzite gravel in channels that trend eastward through the Sheridan Pass and Fish Lake quadrangles into the Wind River Basin. The Yellowstone-Absaroka volcanic province was the third chief source of sediment. Volcanism, as indicated by the fluvial volcanic strata, began in early Eocene and continued into late Eocene time. Deposition of volcaniclastic sediments in the northwestern Wind River Basin began in the Early Eocene (Lost Cabin faunal equivalent). In the eastern Gros Ventre Basin (St. John, 1883), deposition of volcaniclastic sediments began slightly later in early Eocene time than in the Wind River Basin.

Only a part of the stratigraphy of the eastern Gros Ventre Basin can be related to correlatives in the northwestern Wind River Basin. The stratigraphy of the Kisinger Lakes quadrangle and adjacent areas in the northwestern Wind River Basin can be related to established nomenclature of the Du Noir area, which is immediately east of the Kisinger Lakes quadrangle. Tentative correlations of the stratigraphic sequence in the Sheridan Pass, Fish Lake, and Kisinger Lakes quadrangles with fluvial volcaniclastic formations in the East Fork-Wind River area and with the Pitchfork Formation in the eastern Absaroka volcanic province are made. The time stratigraphic framework, based upon age data by Evernden, Savage, Curtis, and James (1964), is presented in figure 3. Only a part of Eocene time is shown. The stratigraphy is summarized in two generalized stratigraphic sections in figure 4.

VOLCANICLASTIC SEQUENCE

The Eocene volcaniclastic rocks are here subdivided into four units (fig. 3): (1) lower volcaniclastic unit, (2) upper volcaniclastic unit, (3) transition zone, and (4) Wiggins Formation. The lower volcaniclastic unit conformably overlies a variegated unit (not discussed) and is disconformably overlain by the upper volcaniclastic unit. The upper unit is disconformably overlain by the transition zone, which is conformable below the Wiggins Formation (not discussed). The lower volcaniclastic unit is of early Eocene age, the upper volcaniclastic unit is in part of early and middle Eocene age, the transition zone is of middle Eocene age.

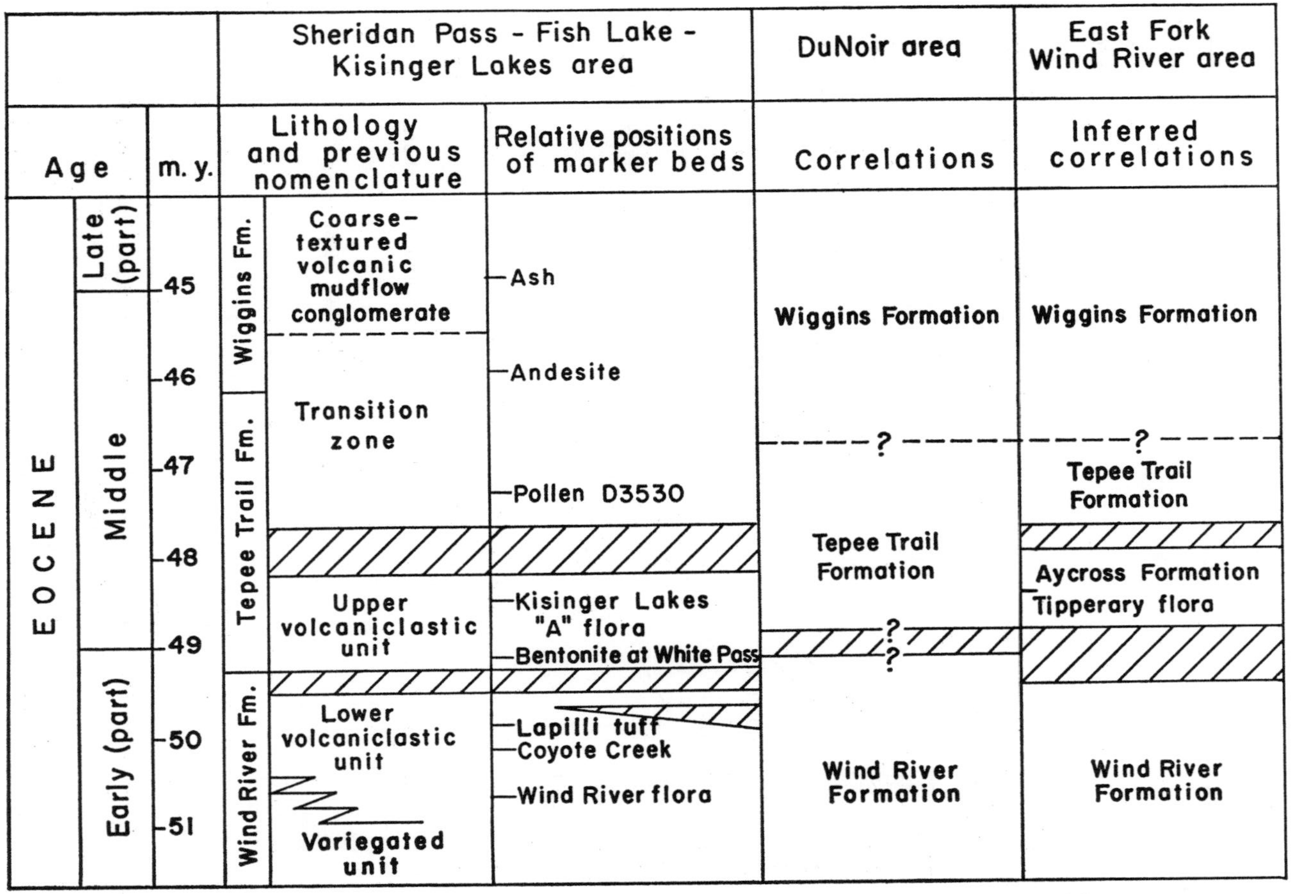

Fig. 3. Ages and lithology of Eocene rocks in the Sheridan Pass-Fish Lake-Kisinger Lakes area and tentative correlations with formations in the Du Noir and East Fork Wind River areas.

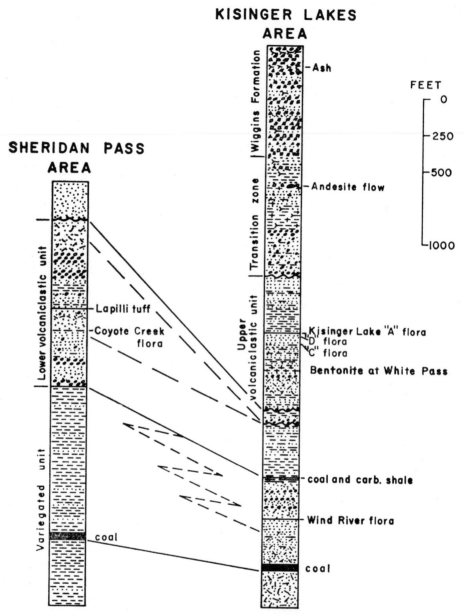

Fig. 4. Generalized stratigraphic sections of Eocene rocks in the Sheridan Pass and Kisinger Lakes areas, showing correlations and positions of marker beds.

There are several diastems in the stratigraphic sections of the fluvial volcaniclastic sequence. Some diastems are relatively insignificant in that only a few feet of strata are absent; some are of greater significance in that several tens of feet of strata are missing. Stratigraphic control on others remains unestablished, and the amount of erosion represented may aggregate a few hundred feet.

The volcaniclastic sequence contains several fossil floras in the northwestern

Wind River and eastern Gros Ventre Basins. These floras occur in various stratigraphic positions and are associated with fluvial and paludal volcanic sediments; there are no indications of volcanic ash fall(s) directly associated with any of the fossil leaf-bearing beds.

LOWER VOLCANICLASTIC UNIT

The lower volcaniclastic unit is present in the eastern Gros Ventre and northwestern Wind River Basins. West of the Continental Divide, in the Sheridan Pass quadrangle, the unit is chiefly exposed in steep landslide scarps. East of the divide, in the Sheridan Pass and Fish Lake quadrangles, the depth of erosion (stream incisement) is much less than west of the divide, and the unit forms rolling to moderately steep hilly terrain which is generally between 9,000 and 10,000 feet in altitude. In the Kisinger Lakes area[1], the unit is exposed in hillsides and landslide scarps from near the level of the Wind River to about 900 feet above stream level.

Lithologically, the lower volcaniclastic unit in the Sheridan Pass quadrangle, which is about 1,200 feet thick, consists chiefly of volcaniclastic granulite[2], sandstone, siltstone, and claystone containing broad eastward-trending channel deposits of massive quartzite cobble conglomerate and mixed arkosic detritus. The granulitic strata are massive unsorted fine-grained volcanic mudflow conglomerates consisting of clay to small pebble-size volcanic fragments and small scattered quartzite pebbles. The sandstone is thin bedded to massive and commonly fine to coarse grained. Sorting is generally poor where the sandstones are arkosic. The sandstone commonly grades upward into siltstone and claystone. Locally, where the strata are relatively highly arkosic there is lateral gradation to finer textured rocks and reddish-colored claystone. The overall color of the fluvial volcanic strata is greenish gray, the quartzite conglomerate is brownish gray, and the more arkosic beds are light gray.

Eastward, in the eastern part of the Fish Lake quadrangle the lower volcaniclastic unit loses much of its fluvial volcanic character, and arkosic strata dominate the sequence. The arkose is coarse textured, thick bedded to massive, and poorly sorted, indicative of fan deposits derived from the Wind River Range. The channels of quartzite conglomerate tend to narrow and trend into the Wind River Basin.

The lower volcaniclastic unit in the Kisinger Lakes area can be divided into two parts. The lower part, about 340 feet thick, consists chiefly of greenish-gray, fine- to coarse-grained volcanic sandstone containing thin interbeds of siltstone, claystone, and quartzite pebble conglomerates. The upper part, about 500 feet thick, consists of greenish-gray claystone and minor beds of red claystone interbedded with siltstone and sandstone. A massive ledge of volcanic pebble conglomerate containing gray to black quartzite pebbles occurs 100 feet below the top of the unit. Except for isolated occurrences in the western and southwestern parts of the Kisinger Lakes quadrangle, the massive channel deposits of quartzite conglomerate are absent in the Kisinger Lakes area.

---

[1] The term Kisinger Lakes area here includes the Kisinger Lakes quadrangle and those parts of adjacent quadrangles west of Du Noir Creek in the northwestern Wind River Basin.

[2] The term granulite is used for clastic grains from 2 to 10 mm.

The upper contact has not been chosen consistently in different areas. Keefer (1957) chose the top of the bentonite bed at White Pass (fig. 3) as the contact between the Wind River and Tepee Trail Formations east of Du Noir Creek. West of Du Noir Creek he did not recognize this bentonite, which is present in the Kisinger Lakes quadrangle and the western part of the Esmond Park quadrangle, and chose the contact at one of the two unconformities below the bentonite (figs. 3, 4). This is about a 300- or 400-foot difference in stratigraphic position. The contact of the lower and upper volcaniclastic units is herein chosen at the upper of these two unconformities because it is distinctive and seems to be more extensive than the lower unconformity. From a time consideration, the top of the bentonite bed at White Pass more closely approximates the hypothetical time line between the early and the middle Eocene than do the unconformities.

STRATIGRAPHIC RELATIONS OF THE FLORAS

One flora, the Wind River flora (USGS Paleobot. loc. nos. D1873, D3535), is present in the lower volcaniclastic unit in the Kisinger Lakes area. It is contained in a microcross-laminated and thinly parallel-laminated claystone, and thick-bedded very fine grained clayey, silty, and biotitic volcanic sandstone about 6 feet thick. The sandstone occurs in the lower part and is gradational into the overlying claystone. The occurrence of fossil leaves in very fine-grained sandstone that grades upward into claystone suggests a floodplain environment. Near-perfect fossil leaves are present in the lower part of the sandstone; however, many are bent and there are numerous leaf fragments. Fossil leaves are sparse between the lower part of the sandstone and the claystone layers. The laminated claystone indicates that swampy conditions prevailed on the floodplain for some time after deposition of the sandstone.

Two key marker beds are present in the lower volcaniclastic unit in the Sheridan Pass area. One of these, containing the Coyote Creek flora, occurs 250 to 450 feet above the base of the unit, and the other, a lapilli tuff, occurs 150 to 160 feet higher in the stratigraphic section (fig. 4). In addition to these marker beds, two local florules occur higher in the section.

The Coyote Creek flora (USGS Paleobot. loc. no. D3534) was originally discovered near the mouth of Coyote Creek in the Fish Lake quadrangle (Rohrer, 1968). The flora occurs in a bed of very fine-grained greenish-gray volcanic sandstone and microcross-laminated and thinly parallel laminated siltstone and shale about 7 feet thick. Mapping in the Sheridan Pass quadrangle (Rohrer, 1969) showed this flora horizon to be extensive, and it has been traced for more than 5 miles through the Sheridan Pass and Fish Lake quadrangles. At some localities there are a few fossil leaves, and at other places the bed is absent owing to postdepositional erosion. In the Devils Basin Creek drainage in the Sheridan Pass quadrangle, the Coyote Creek flora is in a bed about 2 feet thick that consists of volcanic claystone and siltstone. The microcross-laminated and thinly parallel laminated shale and siltstone are indicative of a paludal environment.

The stratigraphic relation between the Coyote Creek and Wind River floras is complicated because of a fault that was active during deposition of the lower volcaniclastic unit. This fault, along Warm Spring Creek in the Fish Lake quadrangle, is upthrown on the north (Wind River Basin) side and resulted in much of the lower volcaniclastic unit being eroded from the Kisinger Lakes quadrangle. The wedge-shaped unconformity within the unit (figs. 3, 4) represents the erosion caused by this fault movement. The time separation of the Wind River and Coyote Creek floras remains unknown, but it seems likely that the separation was brief, probably not more than half a million years.

The lapilli tuff (figs. 3, 4), probably a slurry-flood type of deposit, has no vertical gradation of coarse particles (as much as 1 in. diameter) from the matrix material which has been altered to bentonite and bentonitic claystone. This tuff bed, about 3.5 feet thick, becomes finer textured away from the source area. Radiometric dating of biotite from the lapilli gave a potassium-argon age of about 50 m.y. (J. D. Obradovich, oral commun., April 28, 1971). This date is uncertain because it could not be confirmed.

Two local florules occur stratigraphically higher than the Coyote Creek flora. One is about 250 feet above the Coyote Creek flora (Rohrer, 1968, paleobot. loc. 1). The rock containing this florule is a local pod, about 10 feet wide by 4 feet thick, of thinly laminated claystone and siltstone which is a diastemic remnant of a more extensive bed. The other florule, about 600 feet above the Coyote Creek flora, occurs at several sites along the Continental Divide northwest from the NE¼ Section 36, T 42 N, R 110 W. The enclosing rock is a slightly sandy claystone that appears to have been a fine-textured volcanic ash reworked and redeposited by a stream. The fossil leaves are mostly fragmented. Fossil woods representative of *Plataninium,* Leguminosae, Apocynaceae, and Cupressaceae (R. A. Scott, written commun., June 11, 1970, USGS Paleobot. loc. no. D2065) are common on the surface of white claystone.

Correlations from the west side of the Continental Divide, the Sheridan Pass area into the Kisinger Lakes area, indicate that the volcaniclastic rocks containing the Wind River flora are stratigraphically lower than the lower volcaniclastic unit in the Sheridan Pass area (fig. 2). The lower volcaniclastic unit is thickest (about 1,200 ft.) in the Sheridan Pass area where it conformably overlies variegated strata of early Eocene age (fig. 2). In the Kisinger Lakes area, the unit is much thinner (about 840 ft.) due to unconformities near and at the top of the unit. The unconformity below the top is absent or has not been identified in the Sheridan Pass quadrangle; the unconformity at the top is present in the Fish Lake quadrangle.

The lower volcaniclastic unit is equivalent to part of the Wind River Formation as mapped by Keefer (1957) and Rohrer (1966, 1968). Keefer (1957) informally subdivided the Wind River Formation in the Du Noir area into five sequences. The lower variegated sequence is the lowest of the five and has its westernmost exposure in the west wall of the Du Noir valley. At the Wind River flora site the variegated beds in this sequence are absent because of a facies change to arkosic and volcaniclastic rocks. The beds containing the Wind River flora appear either to overlie directly or to be slightly above the lateral projection of the uppermost beds in Keefer's lower variegated sequence. The thickness of lower Eocene rocks below the Wind River flora ranges from less than 500 feet to more than 1,000 feet near the confluence of Crooked Creek and the Wind River in the northern part of the Warm Spring Mountain quadrangle.

The stratigraphic position of the Wind River flora as related to the Wind River Formation is shown by Keefer (1957:190), and a Lost Cabin faunal age is indicated by vertebrate fossils in associated strata. The uncertain radiometric age of lapilli tuff in the Sheridan Pass quadrangle tends to confirm the early Eocene age.

Deposition of the lower unit volcaniclastic sediments at the Wind River flora site began several tens of feet below the flora beds and continued with brief interruptions and minor incursions by arkosic sediments to the end or nearly the end of early Eocene time.

Upper Volcaniclastic Unit

The upper volcaniclastic unit is present in the Kisinger Lakes area, and a remnant of the lower few hundred feet occurs at Fish Lake Mountain in the Fish Lake quadrangle. In the Kisinger Lakes area the unit is chiefly exposed in a series of landslide scarps. The most complete single outcrop is in the Mud Lake landslide scarp in Section 9, T 43 N, R 109 W.

The upper unit, about 670 to 1,000 feet thick, consists of volcanic sandstone, siltstone, claystone, and volcanic pebble conglomerate. Arkosic sandstone is locally present in the lower part. Sandstone and associated coarse-textured rocks are chiefly limited to the lower 640 feet. Coarse-grained to conglomeratic sandstone is common in the lower 440 feet of the unit, and most of the very fine-grained to medium-grained sandstone is thin bedded to massive and alternates with claystone in the overlying 200 feet. Claystone and bentonitic claystone containing relatively minor interbeds of siltstone and sandstone comprise the upper 320 feet of the unit. The lower 300 feet is greenish gray, the intermediate 400 feet is generally gray to drab brownish gray, and the upper 300 feet is light gray to drab greenish gray.

The upper contact is chosen at a disconformity. Erosion below the disconformity has locally removed nearly 400 feet of the upper volcaniclastic unit.

STRATIGRAPHIC RELATIONS OF THE FLORAS

Fossil leaves occur in several stratigraphic positions, and a key marker, the bentonite at White Pass (Keefer, 1957), is about 260 feet above the base of the unit. Scattered palm fronds occur in the basal strata, and sycamore-type leaves are present in beds below the bentonite bed. The Kisinger Lakes "A" flora is about 530 feet above the base. Two floras occur between the bentonite and the "A" flora (fig. 2). The "C" flora is 90 feet below the "A" flora, and the "D" flora is in a massive sandstone 20 to 50 feet below the "A" flora.

The bentonite at White Pass is 11 feet thick in the NW¼ Section 6, T 42 N, R 107 W, and consists of bentonite and bentonitic claystone (Keefer, 1957:192). In the SE¼ Section 13, T 43 N, R 109 W, this bentonitic zone is grayish tan and 35 feet thick; in the NW¼ Section 9 it is about 80 feet thick and the lower few feet is biotitic. The zone loses much of its bentonitic character 2 miles to the northwest, and a part of the zone is a bentonitic ash. Radiometric dating of biotite from this ash gave a potassium-argon age of 49.3 m.y. (J. D. Obradovich, written commun., August 10, 1970).

The Kisinger Lakes "A" flora (USGS Paleobot. loc. no. D3532A) was originally found in the S½ Section 12 (unsurveyed), T 43 N, R 109 W (Rohrer, 1966), in beds about 5 feet thick. Later search for the flora led to discovery of the same beds in several other sections in the same township. The beds consist chiefly of micro-cross-laminated and finely parallel laminated siltstone and claystone. In Section 5 the flora is absent, but the equivalent strata are much thicker and contain coaly deposits. These strata are briefly described in the following measured section:

	THICKNESS (ft)
TOP	
Siltstone, grayish-black to black, carbonaceous, hard	1.0
Claystone, gray	3.2
Coal, black, clayey, tough (USGS Paleobot. loc. no. D4395c)	0.8
Shale and claystone, gray	7.0
Coal, black, and coaly papery shale (USGS Paleobot. loc. no. D4395b)	1.1
Claystone and siltstone, gray	1.5
Shale, black, coaly (USGS Paleobot. loc. no. D4395a)	1.8
	___
Total thickness	16.4

Rocks in an unmapped area near Barbers Point, NW¼ Section 36 (unsur-veyed), T 44 N, R 110 W, also contain fossil leaves. On the basis of present strati-graphic information, this flora is another occurrence of the Kisinger Lakes "A" flora. Keefer (1957) reported fossil leaves at several localities along Sixmile Creek in the Esmond Park quadrangle. One of these occurrences of fossil leaves may be a lateral equivalent of the "A" flora.

Correlations from the Kisinger Lakes area to the area east of Du Noir Creek indicate that the upper volcaniclastic unit is equivalent to the lower part of the Tepee Trail Formation as mapped by Keefer (1957) and Rohrer (1966). Inasmuch as the Tepee Trail and Aycross Formations (Love, 1939) are undifferentiated in the Du Noir area (Keefer, 1957), their correlation with the upper volcaniclastic unit is uncertain. Radiometric dating of biotite from a tuffaceous facies of the bentonite at White Pass indicates that the lower strata are about 49.3 m.y. old. Radiometric dating of hornblende from an andesite flow in the overlying transition zone indicates an age of about 46.2 m.y. Therefore, the lower part of the upper volcaniclastic unit is of late early Eocene age, and the upper part is of middle Eocene age, and the unit is in part equivalent to the Wind River and Aycross Formations (fig. 1).

TRANSITION ZONE

Strata of the transition zone are present in the Kisinger Lakes quadrangle below the cliffs of the Wiggins Formation, and the zone generally forms steep slopes locally interspersed with ledges. The transition zone is mostly covered by talus and landslide deposits. A single complete outcrop is present in the SW¼ Section 33, T 44 N, R 109 W.

The transition zone, about 700 feet thick, was partly described as the upper part of the Tepee Trail Formation in a preliminary report by Rohrer and Obradovich (1969: fig. 2). The zone is transitional between fine-textured volcaniclastic sand-stone, siltstone, and claystone typical of the upper volcaniclastic unit and coarse-textured volcanic mudflow conglomerates typical of the Wiggins Formation. Thin to thick lenses of coarse-textured volcanic mudflow conglomerate are commonly present in the upper 200 feet of the zone.

Lateral equivalents of the transition zone are coarse textured toward the source area and fine textured toward the basin area. The fine-textured strata locally contain coal or carbonaceous beds. In the Kisinger Lakes quadrangle, a coal bed and a carbonaceous shale bed were sampled for pollen. One sample, D3530 (fig. 3), is about 180 feet above the unconformity at the base of the transition zone, and the other, D3531, is approximately contemporaneous with D3530 but its exact stratigraphic position is uncertain.

The upper contact of the transition zone is chosen at the base of the lowest massive continuous ledge of coarse-textured volcanic mudflow conglomerate typical of the Wiggins Formation. Because strata stratigraphically lower than this ledge become coarser textured toward the source area, the upper contact probably would also be chosen in a lower position toward the source area.

Segments of an andesite lava flow found within a volcanic mudflow conglomerate 600 feet above the base of the transition zone have been radiometrically dated at about 46.2 m.y. (Rohrer and Obradovich, 1969). This date has been confirmed by potassium-argon dating of rocks stratigraphically both higher and lower.

MIDDLE EOCENE CORRELATIONS

The question remains as to how the upper volcaniclastic unit and the transition zone in the northwestern Wind River Basin correlate with the type Tepee Trail, Aycross, and Pitchfork Formations.

The first detailed study of fluvial volcaniclastic strata in the Absaroka Mountains was made by Hay (1956) who defined the Pitchfork Formation. Part of the area where strata of the Pitchfork are present is the area from the South Fork Shoshone River to Ishawooa Mesa (Hay, 1956: 1869). Here Nelson and Pierce (1968) named the coarse breccia facies equivalent of the Pitchfork Formation the Wapiti Formation. It is in this area that the Valley flora, mentioned in another chapter, was found. The Pitchfork was assigned a probable middle Eocene age on the basis of vertebrate and floral evidence. Later, faunal evidence of an early middle Eocene age was reported by Wilson (1963:15). On the basis of these age determinations, the upper volcaniclastic unit in the Kisinger Lakes area correlates at least partly with the Pitchfork Formation.

Tentative correlations of the upper volcaniclastic unit and the transition zone with the Tepee Trail and Aycross Formations are shown in figure 3. There is little doubt that part of the upper volcaniclastic unit is equivalent to part of the Aycross Formation. The Tepee Trail Formation may be equivalent to the transition zone or may be partly represented by the unconformity at the base of this zone. Rohrer and Obradovich (1969) suggested that the contact of the Tepee Trail and Wiggins Formations transgressed time. This is true if the contact is drawn between the coarse volcanic mudflow conglomerates and the fine-grained fluvial facies of these conglomerates. However, Rohrer and Obradovich based their inference in part upon the reported younger ages of these formations in the type localities and elsewhere. New data (fig. 3) indicate that in this area, the Oligocene age of the Wiggins and the late Eocene age of the Tepee Trail are erroneous. [This concludes Mr. Rohrer's contribution to the paper. H.D.M.]

PRESENT CLIMATE AND VEGETATION

The greater part of the area of the western Wind River basin, above an altitude of about 8000 feet, is occupied by the Engelmann spruce-subalpine fir forest cover type (Type 206, forest cover types of North America, Larsen, 1930:42–43). This forest forms dense stands on the gentler slopes of the volcanic soils. The common trees are *Picea engelmannii* and *Abies lasiocarpa* with various amounts of *Pinus flexilis* (limber pine), *Pinus contorta* (lodgepole pine) and *Pseudotsuga menziesii* (Douglas fir). The last three species are more abundant below 8000 feet. *Pinus albicaulis* (whitebark pine) may be seen above 8500 feet. Common shrubs are species of rose, cinquefoil, blueberry, mountain mahogany, service berry, with sagebrush in drier, open spaces. Groves of aspen are scattered around the borders of swampy ground. In the summer, damp, open meadows are glowing with many flowering herbs; larkspur, columbine, geranium, meadow rue, harebell, paintbrush and various composites. These meadows can hardly be exceeded in attractive color during the short, summer season. This forest type is essentially a taiga.

The average annual temperature in this forest is around 32°F or 0°C. The average July temperature is approximately 55°F or 12°C. The growing season is short, probably less than 120 days, over the whole area; above 9000 feet frosts may occur in every month except possibly July. The total annual precipitation on the south-facing slopes of the Absarokas probably exceeds 20 inches, but no data is available in this area. Snowfall is heavy, snow depths ranging from 4 to 12 feet or more in the late winter. At places above 8000 feet in the area of the Wind River headwaters, at the intermingling of the Wind River, Gros Ventre and Absaroka ranges, the snowfall may be as much as 200 inches, and the total precipitation over 35 inches. During the summer, July and August, there are frequent brief thunder showers over the higher ridges. (See Larsen, 1930:651–656.)

There is a different vegetation type in the Tipperary area in the Wilderness quadrangle (Section 18, T 6 N, R 4 W). The locality is at an elevation of 7200 feet on the southern slope of the Owl Creek Mountains. The south-facing slope, the deficient precipitation, and the well-drained, sandy soils combine to eliminate the taiga type of vegetation and to favor a subhumid (or semi-arid) type of grassland vegetation, such as is common in the intermontane basins of the central Rockies. The predominant vegetation cover is a rather sparse growth of grass mixed with sagebrush. In favorable situations there are copses of various low, hardy shrubs; lemonade sumac, service berry, mountain mahogany, snowberry and the like. In places there are scattered individuals of Rocky Mountain red cedar. Along the streams there are groves of aspen, willow leaf cottonwood and willows, with shrubs of chokecherry, rose and snowberry.

The annual precipitation over much of the subhumid, sage and grassland area of the central and eastern Wind River basin is seldom more than 10 inches. For a few selected stations the total precipitation in inches is as follows: Boysen, 8.9; Diversion Dam, 9.5; Lysite, 5.3; Pavillon, 8.7; Riverton, 8.8. The growing season is longer than in the western forested areas, as much as 120 days in the more favored areas around Boysen and Riverton. Snowfall is light, less than 3 feet over most of the central basin. Minimum temperatures reach –30°F in January, and, in the colder areas minima of –40°F have been recorded.

In general the climate of northwestern Wyoming may be described as an extreme continental type, with deficient precipitation, cold winters, low annual temperatures, and relatively great annual ranges of temperature. The fossil flora and its climatic implications, compared with the living vegetation, show how extraordinary have been the changes since the Middle Eocene. The average yearly temperature has decreased by approximately 20°F; absolute minima have declined by 60°F or more; annual rainfall has decreased by 20 to 40 inches; and the elevation has increased by 4000 to 8000 feet or more.

DISPOSITION OF THE SPECIES IN BERRY'S LIST OF
PLANTS FROM THE TIPPERARY LOCALITY, 1930*b*:60
(If no comment follows the name, the species is accepted as valid.)

Equisetum tipperarense Berry
Salvinia preauriculata Berry
Danaea coloradensis Knowlton = Isoetites horridus (Dawson) Brown, not at Tipperary
Dryopteris weedii Knowlton =Thelypteris iddingsii (Knowlton) new combination
Asplenium serraforme Berry = Asplenium eolignitum Berry
Asplenium eolignitum Berry = Asplenium serraforme Berry
Lygodium kaulfussii Heer
Typha sp. Berry
Sparganium antiquum (Newberry) Berry
Sabalites florissanti (Lesquereux) Berry
Sabalites powelii (Newberry) Berry
Geonomites haydenii (Newberry) Knowlton
Musophyllum complicatum Lesquereux
Juglans alkalina Lesquereux, presence doubtful
Juglans occidentalis Newberry = Cedrela schimperi (Lesquereux) new combination
Myrica ludwigii Schimper = Dipteronia wyomingense (Berry), a fragment
Salix sp. Berry = Leguminosites inlustris MacGinitie
Dryophyllum wyomingense Berry = Dipteronia wyomingense (Berry) new combination
Ficus wyomingiana Lesquereux = Apeiba improvisa (Berry) new species
Ficus ungeri Lesquereux = Aleurites fremontensis (Berry) new combination, part, pl. 12, fig. 4;
   pl. 14, fig. 5 = Cedrela schimperi
Ficus mississippiensis Berry = Apeiba improvisa new species
Fagara wyomingensis Berry = Leguminosites wyomingensis (Berry) new combination
Sapindus dentonii Lesquereux, not at Tipperary
Sapindus obtusifolius Lesquereux, from Lenore not Tipperary
Sapindus winchesteri Lesquereux = Cedrela schimperi (Lesquereux) new combination
Negundo fremontensis Berry = Aleurites fremontensis (Berry) new combination
Zizyphus wyomingianus Berry = Populus wyomingiana (Berry) new combination
Ampelopsis tertiaria Lesquereux, type lost, not Ampelopsis
Grewiopsis wyomingensis Berry = Populus wyomingiana (Berry) new combination
Laurus fremontensis Berry = Laurophyllum fremontensis (Berry) new combination
Aralia whitneyi Lesquereux, from Crowheart Butte, not at Tipperary
Aralia brownii Berry = Platanus browni (Berry) new combination
Aralia notata denticulata Berry = Aleurites fremontensis (Berry) new combination
Diospyros mira Berry
Sambucus winchesteri Knowlton, an object of unknown affinity, discarded
Nordenskioldia borealis Heer, not at Tipperary or in the Green River, discarded
Carpites newberryanus Knowlton = Luehea newberryana (Knowlton) new combination
Antholithes anceps Berry, an object of uncertain affinities, unidentifiable ?, discarded
Antholithes brownii Berry, possibly a capsule of Luehea

Antholithus fremontensis Berry, object of unknown affinities
Carpolithus brownii Berry, impression of unidentified fruit
Carpolithus bridgerensis Berry, probably capsule of Luehea
Sporophylls = Isoetites horridus (Dawson) Brown, not at Tipperary
(The objects of uncertain affinity will not be discussed in this paper.)

SYSTEMATIC LIST OF THE FOSSIL FLORA
MEGAFOSSILS OF PTERIDOPHYTA AND SPERMATOPHYTA

SCHIZAEACEA
  Lygodium kaulfussii Heer
PTERIDACEAE
  Acrostichum hesperium Newberry
ASPLENIACEAE
  Asplenium eolignitum Berry
THELYPTERIDACEAE
  Thelypteris iddingsii (Knowlton) MacGinitie
  Thelypteris weedii (Knowlton) MacGinitie
SALVINIACEAE
  Salvinia preauriculata Berry
EQUISETACEAE
  Equisetum tipperarense Berry
TAXODIACEAE
  Glyptostrobus europaeus (Brongniart) Heer
CUPRESSACEAE
  Chamaecyparis sp.
SPARGANIACEAE
  Sparganium antiquum (Newberry) Berry
PALMAE
  Geonomites haydenii (Newberry) Knowlton
  Sabalites florissanti (Lesquereux) Berry
  Sabalites powellii (Newberry) Berry
LEMNACEAE
  Spirodela magna MacGinitie
SALICACEAE
  Populus quintavena MacGinitie
Populus wyomingiana (Berry) MacGinitie
  Salix molesta MacGinitie
JUGLANDACEAE
  Juglans alkalina Lesquereux
FAGACEAE
  Castaneoides aequalita MacGinitie
ULMACEAE
  Zelkova nervosa (Newberry) Brown
ARISTOLOCHIACEAE
  Aristolochia solitaria MacGinitie
LAURACEAE
  Laurophyllum fremontensis (Berry) MacGinitie
  Laurophyllum quotidiana MacGinitie
HAMAMELIDACEAE
  Liquidambar caliarche Cockerell
PLATANACEAE
  Platanus brownii (Berry) MacGinitie
  Platanus intermedia MacGinitie
ROSACEAE
  Prunus nota MacGinitie

LEGUMINOSAE
  Caesalpinites pecorae (Brown) MacGinitie
  Canavalia diuturna MacGinitie
  Leguminosites inlustris MacGinitie
  Leguminosites mira MacGinitie
  Leguminosites occulta MacGinitie
  Leguminosites wyomingensis (Berry) MacGinitie
MELIACEAE
  Cedrela schimperi (Lesquereux) MacGinitie
EUPHORBIACEAE
  Acalypha cirquensis MacGinitie
  Aleurites fremontensis (Berry) MacGinitie
AQUIFOLIACEAE
  Ilex sclera MacGinitie
ACERACEAE
  Dipteronia wyomingense (Berry) MacGinitie
SAPINDACEAE
  Serjania rara MacGinitie
TILIACEAE
  Apeiba improvisa MacGinitie
  Luehea newberryana (Knowlton) MacGinitie
STERCULIACEAE
  Sterculia subtilis MacGinitie
DILLENIACEAE
  Saurauia propia MacGinitie
MYRTACEAE
  Eugenia americana (Knowlton) MacGinitie
ARALIACEAE
  Dendropanax latens MacGinitie
  Schefflera insolita MacGinitie
CORNACEAE
  Cornus sp. MacGinitie
EBENACEAE
  Diospyros mira Berry
SYMPLOCACEAE
  Symplocos incondita MacGinitie
INCERTAE SEDIS
  Alafructus lineatulus (Cockerell) MacGinitie
  Ampelopsis tertiaria Lesquereux
  Carpites araliodes MacGinitie
  cf. Cissus sp. MacGinitie
  Proteaciphyllum minuta MacGinitie

ECOLOGY OF THE FOSSIL FLORA

INTRODUCTION

In the time interval since the Kisinger Lakes-Tipperary flora flourished, an interval of some 45 to 50 million years, there have been changes in the morphology and adaptations of the genera and species found in the fossil flora. In general the processes of evolution which cause these changes are continuing; they do not cease. The fossils were identified by comparisons with the leaves and fruits of living genera and species. Obviously, because of evolutionary changes, the older a Tertiary flora the less marked the correspondences between living and fossil forms. As we ascend the geologic column in time the similarities between living and fossil species become more marked, and, in the case of Pliocene floras for example, plant megafossils can be identified with living forms with considerable accuracy. Thus the difficulties of identification naturally increase with the age of the flora, and the morphology and adaptations of the most similar living species become more of an approximation to those of the fossil plants. This difficulty was illustrated by Reid and Chandler (1933) during their work on the London Clay flora, in which they felt it was necessary to use many new generic terms rather than to indicate a degree of relationship which did not seem justified. In the Wyoming fossil flora there are all degrees of resemblances to the foliage and fruits of living species. In some cases the resemblances are so close as to convince the researcher of a close relationship with a living species; in others the morphological similarities are much less clear. Thus I have practically abandoned the generic identification of the legumes and laurels in the Eocene floras and have used the terms *Laurophyllum* and *Leguminosites*. This philosophy can be carried too far and might lead to an excuse for less intensive work of comparison. Some of the genera exhibit a rather striking stability of gross, foliar morphology with the passage of time. A relatively efficient stage in leaf structure seems to have been reached and further evolutionary changes have become almost imperceptible. In the fossil flora this is illustrated by the genera *Acrostichum, Canavalia, Dipteronia, Liquidambar, Lygodium, Prunus, Sterculia* and *Zelkova*, whose fossil forms show close correspondences with the foliage of the living genera. The foliage of *Platanus* and *Populus,* among others, in contrast has shown marked changes with respect to time. These remarks are to emphasize that comparisons and correspondences between living genera and species and the fossil taxons must necessarily be approximations of varying degrees, in considering a flora as old as the Paleogene. In some other instances, in addition to *Leguminosites* and *Laurophyllum,* it might have been nearer the truth if I had used a new generic epithet. With the possible exceptions of the ferns, all the species appear to be extinct.

The paleobotanist, engaged in research on floras, in his striving to identify the fossil remains, tends to overemphasize resemblances to living species, and thus may give the impression of the stability of foliar and pollen morphology, which is not justified. A study of paleobotanical publications and fossil collections indicates that the general foliar morphology in many of the woody angiosperm families was well established by the Middle Eocene, but there have been evolutionary changes of varying degrees in most taxa since that time. This is illustrated by the foliage of the families Fagaceae, Salicaceae, Betulaceae and Platanaceae. An extraordinary diversity of foliar morphology occurs in these families throughout the Tertiary, although the family characteristics are clear enough to warrant generic assignments. The changes cannot be adequately shown until the evolution of each family is independently monographed.

Since the worker in Tertiary paleobotany must necessarily obtain the greater part of the relationships between fossil and living species from the study of herbarium material, it is necessary to rely on the annotations of herbarium specimens. These annotations in some instances may be considerably outdated in terms of modern taxonomy, and it is not always possible to determine the presently accepted species names. Thus some apparent inconsistencies in designating the most similar living species may occur.

Fossil Species and Similar Living Species

In table 1 which shows the distribution of similar living species, there must exist a certain amount of subjectivity. Also it is obvious that no worker can examine specimens of all the woody plants of any particular genus in many cases, and thus he may overlook a more similar species than the one indicated. All this depends on available time and the type and number of herbaria visited, in addition to familiarity with living forests showing similarities to the composition of the fossil flora. Nevertheless this kind of table represents a large amount of intensive work and is of indispensable value in determining the ecology of a Tertiary flora.

The term flora is essentially a taxonomic concept, whereas the term vegetation refers to the morphological aspect of the plants in any particular area. The term vegetation implies the adaptation of the flora to the particular climatic and edaphic factors present. To illustrate, Wang, in his Forests of China (1961:1) calls attention to the fact that in China the woodland extends in an unbroken succession of forest communities from north to south: (1) boreal coniferous forest, (2) mixed northern hardwood forest, (3) temperate, deciduous, broad-leafed forest, (4) mixed mesophytic forest, (5) evergreen, sclerophyllous forest, (6) tropical rain forest. These are vegetation types. The vegetation type depends on the floristic composition, and floras change with time, but vegetation types may persist, depending on the recurrence of particular conditions. It is easy to confuse the concepts of flora and vegetation, but they are distinct, and treating them as synonyms obscures many important aspects of paleobotany and ecology.

Any habitable land area must have been covered by vegetation since the advent of land plants (Mason, 1947:202). The flora of any particular area must continually change even if the same vegetation type can persist in the area for several million years. Floras change by and through genetic modifications (mutations, recombinations, etc.), selection, dispersal and extinction. These processes are inexorable, and, under natural conditions they are not discontinuous; they never stop. Speciation cannot be stopped.

A fossil flora is a momentary thing. It preserves for us the flora and vegetation type for a small interval of time. In a sense a fossil flora represents a stopping of the geologic clock; and one page in the history of the continuing change in the plant cover is preserved for us.

Thus, in attempting to reconstruct the paleoecology of an area in terms of its Eocene fossil plants, the conclusion seems unavoidable, that, because of the processes of evolution, stimulated at times by topographic changes, it is not possible that any particular *floristic* group will be repeated as time passes. However, the vegetational type, the particular group of plant forms adapted to a particular habitat, may be repeated in time, although the floristic composition changes. The researcher in paleobotany attempts to reconstruct the type of vegetation, but his evidence must stem from floristics, the taxonomic group identified by comparison with living entities. The adaptations of any one species (or genus) may have changed, but if the habitat requirements of the major number of species in the flora points consistently toward one vegetation type, a satisfactory reconstruction of the paleoecology for the past time interval in the area can be made. Thus the accuracy of the ecologic reconstruction depends on accurate identifications, and

FOSSIL SPECIES	SIMILAR LIVING SPECIES	HABITAT LIVING SPECIES						
		1	2	3	4	5	6	7
ACALYPHA CIRQUENSIS	ACALYPHA EROSA	X	X			X		
ACROSTICHUM HESPERIUM	ACROSTICHUM AUREUM					X		
ALEURITES FREMONTENSIS	ALEURITES MOLUCCANA	X	X				X	
AMPELOPSIS TERTIARIA	PLATYCARYA ? sp.				X			
APEIBA IMPROVISUS	APEIBA ASPERA					X		
ARISTOLOCHIA SOLITARIA	ARISTOLOCHIA ANGUICIDA	X		?		X		
ASPLENIUM EOLIGNITUM	ASPLENIUM SERRA						X	
CANAVALIA DIUTURNA	CANAVALIA SEPTENTRIONALIS	X				X		
CARPITES ARALIOIDES	DENDROPANAX sp.,					X		
CASTANEOIDES AEQUALITA	CASTANEA SEGUINII	X	X		X			
CEDRELA SCHIMPERI	CEDRELA ANGUSTIFOLIA	X				X		
CHAMAECYPARIS sp.	CHAMAECYPARIS sp.			X				
cf. CISSUS sp.	EXTINCT							X
CORNUS sp.	CORNUS sp.	X		X	X			
DENDROPANAX LATENS	DENDROPANAX ARBOREUS	X				X		
DIOSPYROS MIRA	DIOSPYROS VIRGINIANA	X						
DIPTERONIA WYOMINGENSE	DIPTERONIA SINENSE	X			X			
EQUISETUM TIPPERARENSE	EQUISETUM GIGANTEUM					X		
EUGENIA AMERICANA	EUGENIA (JAMBOS) JAMBOS					X		
GEONOMITES HAYDENII	GEONOMA spp.					X		
GLYPTOSTROBUS EUROPAEUS	GLYPTOSTROBUS HETEROPHYLLUS				X			
ILEX SCLERA	ILEX ROTUNDA				X			
JUGLANS ALKALINA	JUGLANS NIGRA	X	X	X				
LAUROPHYLLUM FREMONTENSIS	MACHILUS VELUTINA, PHOEBE sp.				X		X	
LAUROPHYLLUM QUOTIDIANA	PERSEA CINERASCENS					X		
LEGUMINOSITES INLUSTRIS	DIOCLEA REFLEXA					X		
LEGUMINOSITES MIRA	LONCHOCARPUS spp.					X		
LEGUMINOSITES OCCULTA	CASSIA sp.					X		
LEGUMINOSITES WYOMINGENSIS	DERRIS TRIFOLIATA						X	
LIQUIDAMBAR CALLARCHE	LIQUIDAMBAR STYRACIFLUA	X	X					
LUEHEA NEWBERRYANA	LUEHEA SPECIOSA	X				X		
LYGODIUM KAULFUSSII	LYGODIUM PALMATUM	X						
PLATANUS BROWNI	PLATANUS RACEMOSA	X						X
PLATANUS INTERMEDIA	PLATANUS LINDENIANA	X						?
POPULUS QUINTAVENA	POPULUS LASIOCARPA	X		X				
POPULUS WYOMINGIANA	POPULUS DIMORPHA	X			X			
PROTEACIPHYLLUM MINUTA	HAKEA spp.							X
PRUNUS NOTA	PRUNUS CAPULI	X	X			X		
SABALITES FLORISSANTI	SABAL ? sp.							
SABALITES POWELLII	SABAL ? sp.							
SALIX MOLESTA	SALIX BONPLANDIANA	X	X					
SALVINIA PREAURICULATA								
SAURAUJIA PROPIA	SAURAUIA BELIZENSIS	X	X					
SCHEFFLERA INSOLITA	SCHEFFLERA HETEROPHYLLA	X					X	
SERJANIA RARA	SERJANIA RACEMOSA	X						
SPARGANIUM ANTIQUUM	SPARGANIUM EURYCARPUM			X				
SPIRODELLA MAGNA	SPIRODELLA POLYRRHIZA							X
STERCULIA SUBTILIS	STERCULIA JAVANICA	X	X				X	
SYMPLOCOS INCONDITA	SYMPLOCOS TINCTORIA	X	X	X				
THELYPTERIS IDDINGSII	THELYPTERIS DENTATUS						X	
THELYPTERIS WEEDII	THELYPTERIS spp.					X		
ZELKOVA NERVOSA	ZELKOVA SERRATA	X	X		X			
		25	19	7-8	9	25	13	4-5

Table 1. Living species similar to the fossil species and their habitats, with two leaf characters. 1. Deciduous. 2. Non-entire margin. 3. Southern Appalachians. 4. Warm temperate to tropical Eastern China and Japan. 5. Southwestern Mexico to northern South America, tropical semi-deciduous forest. 6. Southern China and Indo China, tropical semi-deciduous. 7. Extinct or of doubtful status.

accurate interpretations of the habitat requirements of each one of the most similar living species. This floristic approach can be supplemented by certain physical characteristics of the fossil plants and of the enclosing sediments—analysis of leaf margins, study of fossil wood morphology, chemistry of the sediments and the like.

The Kisinger Lakes-Tipperary flora of the northwestern Wind River Basin, in terms of the most similar living species, exhibits no outstanding resemblance or markedly close relationship to any existing floristic group. However, as will be shown later, there is a significant relationship to the semi-deciduous forest of southwestern Mexico, the bosque tropical subdeciduo. The floristic resemblance and the physical characters of the leaves indicate that the fossil flora had the attributes of a tropical or near-tropical, semi-deciduous forest. This vegetational type is common in the tropics today.

The taxonomic list of the fossil flora shows 5 ferns, 1 horsetail, 2 gymnosperms, and 44 angiosperms. The angiosperms include 3 palms and 2 herbs, *Sparganium* and *Spirodella*. Five of the angiosperm species are of doubtful status, leaving 36 well-defined woody dicots. Of the 36 species of woody dicots, 3 are vines and the others trees, or in a few cases possibly treelike shrubs. Thus the fossil megaflora is essentially a tree flora.

The 11 most abundant species in the fossil flora, approximately in the order of numerical abundance are: *Platanus brownii, Lygodium kaulfussii, Canavalia diuturna, Sabalites* spp., *Cedrela schimperi, Symplocos incondita, Populus wyomingiana, Ilex sclera, Acalypha cirquensis, Dendropanax latens* and *Aleurites fremontensis*. The flora might be described or characterized as a *Cedrela-Platanus-Populus-Symplocos* association with abundant palms and *Lygodium*. This is clearly a floodplain flora in a warm climate. Approximately 20 of the correlative living species are found in eastern and southern Asia, and 30 in America from the southern Appalachians south into Mexico and adjacent tropical America. Of the 11 most abundant species (10 or more specimens collected), 8 are most similar to species of living American genera: *Acalypha, Canavalis, Cedrela, Dendropanax, Platanus, Populus, Sabalites, Symplocos*. Only three of the most abundant species are clearly related to species of Asiatic genera: *Aleurites, Ilex* and *Lygodium*. The flora is thus distinctly of American aspect in terms of present-day vegetation. This is in contrast to the older Wind River flora of Early Eocene age (under study by Hickey and MacGinitie) which has a more marked Asiatic aspect.

LEAF MARGIN CHARACTERS

Of the 36 dicots, 20 have non-entire leaves, approximately 55 percent. In terms of the similar living species 22 of the 36 dicots were *deciduous,* or 61 percent. The proportion of *entire* leaves in the flora, 45 percent, is intermediate between that of the Simla Hills, India, and the southeastern United States, 61 and 43 percent respectively (Bailey and Sinnott, 1915, 1916). It is about the same as the subtropical forest of Taiwan at altitudes between 500 and 1000 meters, that is 47 percent (Wolfe, 1971:34). The percentage of entire leaves in the fossil flora would tend to indicate a "C" climate rather than an "A" climate. There appears to be a contradiction here between the temperature indications of the proportion of entire leaves and the floristic composition of the flora. There is a relatively large group of genera, represented in the list of similar species, which is now essentially tropical in adaptation: *Acalypha, Acrostichum, Apeiba, Canavalia, Cedrela, Eugenia, Dendropanax, Machilus-Persea,* legumes of tropical affinities, *Luehea, Saurauia, Schefflera, Serjania* and *Sterculia.* Seventeen of the 36 dicot species are now tropical.

TABLE 2

LEAF CHARACTERS OF SOME LIVING FLORAS

	Percent entire	Percent non-entire	Percent deciduous
Simla Hills	61	39	
Semi-deciduous, Mexico.	59	41	60
Taiwan, middle altitude, 500–1000 m.	47	53	
Kisinger Lakes-Tipperary	45	55	61
Southeast United States	43	57	
Mesophyll montane Mexico	40	60	55
Alabama forest	23	77	81

There are two other factors which must be taken into consideration in interpreting the climatic significance of the percentage of entire leaves among the woody dicots. There is some evidence that the percentage of entire margins is less in floodplain and lakeside assemblages (Wolfe, 1971:36). The fossil flora under consideration is essentially a floodplain flora and this would tend to lower the percentage of entire margins. Also, there is a direct relation between non-entire margins and the deciduous habit. A survey of the woody, dicot flora of Alabama (Harper, 1928: pt. 2) showed 77 percent non-entire and 81 percent deciduous. The large percentage of non-entire leaves in the woody flora of temperate, humid, North America, is the result in part of the dominance of such plant families as Aceraceae, Betulaceae, Fagaceae, Juglandaceae, Rosaceae and Salicaceae (see Wolfe, 1971:38–39). Since twenty-two or 61 percent of the fossil, woody dicot species were probably deciduous, this relatively large proportion of deciduous species would tend to reduce the percentage of entire margined species. Thus the deciduous character of the flora, together with its floodplain habitat, would tend to increase the percentage of non-entire leaves, and in terms of the Bailey and Sinnott data, would cause the flora to appear more temperate than it really was.

The table of leaf margin percentages for some Teritary floras of North America given by Wolfe and Hopkins (1967:70–71) is obtained from floras which are either lowland coastal or which, if inland, were probably essentially under the influence of Pacific maritime air masses. The exceptions are the Paleocene floras: Denver and Fort Union. The Tertiary floras of the central Rocky Mountains grew in a continental climate, and from the late Early Eocene on these floras comprised a significant proportion of deciduous species. This suggests that paleotemperature estimates based on leaf morphology might be somewhat different for the continental floras than for those from the West Coast. What is needed is a thorough survey of the American tropical deciduous forests from Mexico to and including South America with respect to leaf morphology. The results of such a survey might solve the apparent contradictions between the temperature indications of leaf morphology and the taxonomic composition of the Green River and Kisinger Lakes floras (MacGinitie, 1969:45).

It is apparent from the Florissant, Green River and Kisinger Lakes studies, that the Bailey and Sinnot values apply to floras which existed under uniformly distributed rainfall during the year. Leaf statistics for the tropical deciduous or semi-deciduous floras must be treated separately; these show a larger percentage of non-entire margins than do tropical floras which live in regions characterized by rainfall distributed fairly evenly throughout the year.

Wolfe (1971:35) says, "There is clearly some adaptive significance between the type of leaf margin and climate, although the exact physiological relationships are not known." The correlation between the deciduous habit and non-entire leaf margins is so marked that these two factors must derive from the same environmental stimulus. Axelrod (1966) has assembled evidence to indicate that deciduous hardwoods originated in lower latitudes in response to seasonal drought. It is clear that the deciduous habit is a response to seasonal, physiological dryness, whether caused by lack of water in the liquid state (freezing) or by actual lack of water (drought). Although the relationship between non-entire margins and a dry season must be complex, the conclusion seems unavoidable that the development of non-entire margins is at least one common response to seasonal drought.

In the matter of leaf sizes a rough estimate can be made in terms of total length. (See Webb, 1959.) The classification used here is: under 7.5 cm., microphyll; 7.5 cm. to 12.5 cm., notophyll; over 12.5 cm., mesophyll. Using 43 species of dicots from the Kisinger Lakes flora, a few of which are not yet identified, the size range is as follows: microphyll, 25 percent; notophyll, 35 percent; mesophyll, 40 percent. This resembles the semi-evergreen mesophyll vine forest of Webb which shows: microphyll, 30 percent; notophyll, 40 percent; mesophyll, 30 percent. The Wyoming flora shows a considerably larger percentage of moderately large leaves than the Australian flora. This may have been caused by a higher rainfall and, perhaps, a longer rainy season. Also, it must be emphasized again that the fossil flora is biased; it is not a good sampling of the general vegetation but is much weighted toward floodplain vegetation. The percentage of vines cannot be accurately given; the following were probably vines: *"Ampelopsis," Aristolochia, Canavalia, Serjania,* and possibly *Leguminosites wyomingensis.* This represents about 14 percent of the woody, dicot species. The *Lygodium,* of course, is a vine.

One of the striking characteristics of tropical climates, that is between the parallels of 30°S and 30°N, roughly, is the wide expanse of Köppens "w" climates with a dry season in the winter half year (see Bruzon and Carton, 1930). About the only significant exceptions are found in the true equatorial climates, on east coasts in the trade wind belts, and in certain west coast areas in equatorial regions. The rainy east coast climates are illustrated by the Gulf coastal strips of southern Mexico and Central America, southeast Brazil and eastern Australia. The rainy west coast climates are found on the coasts of Liberia and Nigeria in Africa and the west coast of Colombia in South America. The "w" climates prevail over vast areas in Mexico, South America, southern Asia, and Africa. It might be said that, with the exceptions noted above, the typical climate of the tropics is a winter-dry climate. It is in these areas that the great expanses of modern savanna vegetation is found. The forests are characteristically deciduous or semi-deciduous. The same thing must have been true of these climates during the Tertiary.

With the warmer world climate of the Eocene and the lower temperature gradient north-south, the zone of traveling cyclones, which is so well developed today during the winter half-year in temperate latitudes, must have been much weaker and of less intensity than it is at present. This would tend to produce drier winters, especially toward the centers of the continents. The time of maximum precipitation would correspond with the time of maximum surface heating during the summer half-year just as it does over much of Mexico and Central America today.

The evidence for a dry season is strong in the case of the Middle Eocene Green River flora of northern Colorado and Utah (MacGinitie, 1969:46–50). This flora probably lived in a region where the climate bordered on the subhumid. The Kisinger Lakes-Tipperary flora experienced a higher annual precipitation, but the percentage of deciduous, woody angiosperms indicates that there was a rather well-marked dry season.

DISTRIBUTION OF CORRELATIVE LIVING SPECIES

Considering the fossil flora as a whole, approximately 20 of the correlative living species grow today in the American tropics. As far as can be determined none of these is found far north of 25°N latitude. About 12 similar species are found in the tropical areas of southern China and Indo-China. Fifteen to 16 species inhabit the southern Appalachians or warm temperate China and southern Japan. It should be noted, that, in the case of the genera which today extend into temperate areas, the correlative living species is characteristic of the warmer aspects of the habitat area. The genus *Acalypha*, for example, contains close to 400 species, of which some are herbaceous or shrubby and extend into the southern United States. But the most similar species found is *A. erosa*, a small tree with leaves of firm texture, which is found only in the tropics. The genus *Ilex* comprises about 180 species, a few of which are temperate to cool temperate in adaptation, but *Ilex rotunda*, which most resembles the fossil *Ilex*, is found in warm temperate to tropical southeast Asia. In this connection we can recall the characteristically tropical distribution of the genera *Persea*, *Sabal* and *Diospyros*, but which have been able to evolve a few species which extend northward into the United States where comparatively severe freezes occur. How old are these "temperate" species of tropical genera? Thus the aspect of the group of similar living species is warm temperate to tropical, with a leaning more toward the tropical (see Table 1).

The following correlative living species are essentially tropical in their modern distribution, whether American or Asiatic:

Acalypha erosa	Geonoma sp.
Acrostichum aureum	Lonchocarpus sp.
Aleurites moluccana	Luehea speciosa
Apeiba aspera	Sabal spp.
Canavalia spetentrionalis	Saurauria belizensis
Cedrela angustifolia	Schefflera heterophylla
Dendropanax arboreus	Serjania racemosa
Dioclea reflexa	Sterculia javanica
Eugenia jambos	Thelypteris spp.

This group of 18 forms is complemented by several tropical families (and genera) found in the pollen flora: *Trema* (Ulmaceae), *Bombax* and *Bernoullia* (Bombacaceae), *Cedrela* (Meliaceae), *Bauhinia* (Leguminosae), cf. *Anaclosa* (Olacaceae) and *Triumfetta* and *Luehea* (Tiliaceae). (The pollen flora is treated in the chapter by Dr. Leopold.) Thus there is a marked, tropical element in the group of correlative living species. However there is also a group of about a dozen species whose adaptations are essentially to temperate climates.

One of the common difficulties faced by a student of the Paleogene floras is reconciling the association of apparently temperate and apparently tropical species; the mixture of species whose living representatives inhabit both "temperate" (Köppen's C climates) and tropical (Köppen's A climates) (Köppen, 1931).

Of the group whose living representatives are essentially temperate in their adaptations, there are some which are rare in the fossil flora, 5 or fewer specimens collected; and these may have been strays from some upland area. Among these are: *Castaneoides, Chamaecyparis, Juglans, Liquidambar, Populus quintavena* and *Prunus*. A few other species among the "temperate" group are relatively abundant as fossils and these species must have been members of the floodplain flora; *Dipteronia, Populus wyomingiana, Salix* and *Symplocos*. *Dipteronia* does not appear to extend its habitat into the tropics at present, but does inhabit typically warm temperate climates. The *Populus wyomingiana* exhibits a venation like that of the living *Populus dimorpha* which lives on the west coast of Mexico north of the tropic but in a climate, due to its coastal position, which is essentially tropical. There is no clear evidence that this form of *Populus*, when found in a Paleogene flora, should be considered of temperate habitat. *Salix bonplandiana* and *Symplocos prionophylla* are found in the bosque mesofila de montaña which extends as low as 900 meters above sea level along protected canyon sides of the Sierra Madre Occidenal in Nayarit and western Jalisco, Mexico. This is in the transition between the higher oak-pine forest and the tropical semi-deciduous forest below. The zone where freezing occurs is above 1200 meters in this area. The genus *Symplocos* has many species in Taiwan and neighboring southeast Asia which are in part of tropical adaptation. Since the fossil species are all clearly extinct it is safe to say that we can never be absolutely certain concerning their adaptations in the Eocene. What weight are we to give these seemingly temperate species in determining the ecology of an Eocene flora?

There are two ways of explaining this common, early Tertiary mixture of tropical and temperate (at present) species and genera. The first is to assume a climate similar to that of certain areas in tropical highlands today, such as that at Huachinango or Orizaba, Mexico. Some of these localities have an annual average temperature below 70°F (21°C) but are able to support many "tropical" species due to the high minima, the temperateness. I have designated this as the Orizaban Subtropical climate (MacGinitie, 1969:42–43). This climatic type may be a fairly accurate delineation of the climate under which the Early Eocene Chalk Bluffs flora of California existed (MacGinitie, 1941:62–78) but it is less satisfactory in the case of the Green River flora or the early Middle Eocene, Aycross flora under discussion.

The second approach is to deemphasize the seemingly temperate species as such, and to suppose that this group was more tropical in adaptation at the time. The present adaptation to cool climates with severe winters, of the genera *Dipteronia, Populus, Salix, Symplocos, Carya* and the like may have developed subsequent to the Eocene. Since the world in general has, at present, a cooler and more intemperate climate than that of the earlier Tertiary and Cretaceous, we can say, rather confidently, that trees adapted to the modern temperate or cool temperate climates of middle latitudes are probably more advanced or evolved than their relatives in the tropics. In this connection it appears that temperate species of mainly tropical genera such as *Diospyros, Ilex, Magnolia* and *Persea* may be rather late products of evolution.

Beginning in the late Paleocene there seems to have been a long time interval during which earth climates were warm and equable. There is evidence of a cool period in the latest Eocene but the marked climatic break (at least in the northern hemisphere) was in the Middle Oligocene (Wolfe, 1969:68, 84–88, 1971: 50–51; Devereux, 1967; Jenkins, 1968; Nemejč, 1964). The Oligocene climatic cooling is clearly seen in Professor Mai's diagram of vegetational changes during the Tertiary in middle and western Europe (1964:149), which shows a decline in tropical elements from about 56 percent in the Middle Eocene to about 22 percent in the Upper Oligocene. The diagram also shows the warm fluctuations in the Miocene. Previous to this time (later Oligocene) lowland floras at middle latitudes have a marked tropical aspect. After the Middle Oligocene there is a definite change, both in America and Europe, to floras of more temperate aspect. In this period of climatic change there must have been a premium on adaptations to cooler and less favorable conditions, and thus on the development of new species with new tolerances. The middle Oligocene climatic break must have stimulated the temperature adaptations of a great many of the trees of middle and northern latitudes. At least some of the oaks, elms, Betulaceae, Rosaceae and other woody plants of temperate latitudes may not have had their present adaptations in the Eocene.

Sycamore (*Platanus*) leaves, or at least those of close relatives of the Sycamore, are fairly abundant in the later Cretaceous (tropical or near tropical) floras of North America, but remains of *Salix* or *Populus* are uncommon or absent, judging from the illustrations and descriptions by Berry, Knowlton, Lesquereux and others. Berry illustrates an apparently valid fossil of *Salix* in his Upper Cretaceous flora of South Carolina (1914:pl. 8, figs. 11, 12). *Salix lancensis* Berry is relatively abundant at several localities in the Lance Formation of Late Cretaceous age on Lance Creek in eastern Wyoming. This appears to be well-authenticated foliage of *Salix* and one of the earliest occurrences of the genus. (Dorf, 1942:134.) *Populus* seems to have diverged from *Salix* (or vice versa) in the latest Cretaceous or early Paleocene. There is no cogent reason for believing that the genera *Salix* and *Populus* were good indicators of temperate conditions when we find them in the Eocene, for example.

The problem of mixed temperate and tropical species was discussed in the paper on the Kilgore flora of Mio-Pliocene age (MacGinitie, 1962:89–92), and it was there suggested that climatic tolerances of certain species might have changed since the late Miocene. This would be all the more true of older floras. By this second approach we avoid some of the *apparent* contradictions in the adaptations of the various elements of a fossil flora. Gray (1960) has discussed the significance of temperate pollen genera in the Claiborne flora, and considers that these may have been transported from some bordering upland area.

It is not to be inferred from the above discussion that I believe there were no genera or species of the Paleogene angiosperms adapted to temperate conditions. Tertiary pollen of *Alnus* in tropical areas seems clearly connected with upland habitats (Germeraad, Hopping and Muller, 1968:273). During the long history of angiosperm evolution in the Cretaceous and Paleocene, adaptations to cool upland areas must have developed in many taxa. The idea stressed is that such species of *Populus* and *Salix,* for example, as those which appear in the Wyoming flora should not be taken as unequivocal evidence of temperate conditions at the time. In this connection palynological evidence indicates that *Alnus* as a genus, is older than *Populus* or *Salix,* since characteristic alder pollen occurs in the later Cretaceous.

The paleobotanist, working with ferns of the past, acquires the impression that evolution of the group has been extremely slow during the Tertiary, even imperceptible in some genera. However, it will take much more intensive research to satisfactorily demonstrate this. Many of the fern genera may date from the middle Mesozoic. If this is true, then the adaptations of fern species probably have changed little since the early Tertiary. Thus, in dealing with the climatic implications of a group of fossil fern species, there should probably be no serious problems concerning evolutionary changes in tolerances and adaptations since the Eocene. This narrows the problem down to correct identifications of the fossil material. The 4 fern genera of the Kisinger Lakes flora all possess definitive leaf venation, and the accuracy of the identifications is beyond all reasonable doubt. The genus *Acrostichum* is essentially tropical in its present distribution, although its range extends northward to central Florida where frosts occur. Although the genus *Asplenium* reaches its greatest development in the tropics, it is not confined to tropical climates. Several species are found in temperate regions. The most similar living species is a tropical form. The genus *Lygodium* is found in both tropical and temperate areas. One species is found as far north as latitude 43°N on the eastern seaboard of the United States. The most similar living species inhabits warm temperate Japan. The genus *Thelypteris* is characteristically warm temperate to tropical in its present distribution, but some species are by no means confined to warm climates. Dense thickets of *Acrostichum* and *Thelypteris* grow with various "cane" species around the borders of the mangrove swamps of coastal western Mexico, from Mazatlan south.

Thus we are faced with the usual "temperate-tropical ambiguity" that we find in dealing with the fossil woody angiosperms of the Paleogene. It seems clear that climatic implications of fern genera can give only approximations to fossil climates. It is extremely difficult or impossible to establish exact correspondences between the fossil forms and living species. Some of the apparent "climatic ambiguities" may stem from this fact.

The following living correlative species are members of the semi-deciduous, tropical forest in southwestern Mexico, in the area between San Blas and Tepic-

Puerto Vallarta, at altitudes of from 0 to 3000 feet on the western slopes of the Sierra Madre Occidental:

Acalypha schiediana
Acrostichum danaefolium
Apeiba aspera
Aristolochia anguicida
Canavalia villosa, brasiliensis
Cornus sp.
Cedrela sp.
Dendropanax arboreus
Eugenia sp.

Persea cf. cinerascens
Lonchocarpus sp.
Luehea candida
Lygodium sp.
Sabal sp.
Salix chilensis
Serjania racemosa
Spirodella polyrrhiza
Thelypteris sp.

Other similar living species are found in the lower borders of the pine-oak forest (bosque mesofila de montaña) at altitudes of between 3000 and 4000 feet:

Canavalia sp.
Cornus disciflora
Dendropanax arboreus
Eugenia sp.
Ilex brandegeana

Persea sp.
Salix bonplandiana
Saurauia serrata
Sabal sp.
Symplocos prionophylla

Around the borders of the mangrove swamps are dense thickets of *Acrostichum* and *Thelypteris*. *Aristolochia* and *Serjania* (or Paullinia) are common vines in the transition between this forest and the dryer deciduous forest (bosque tropical deciduo).

The fossil flora lacks such characteristic genera of the tropical semi-deciduous forest in this area as:

Brosimum
Bursera
Castilla
Ceiba
Cecropia
Celtis
Cochlospermum

Enterolobium
Ficus
Guarea
Orbignya
Piper
Randia
Tabebuia

Orbignya, the giant feather palm, in places becomes a dominant, and three species of large figs are abundant. *Brosimum, Guarea, Cecropia, Castilla,* and *Ceiba* are common everywhere in the semi-deciduous, tropical forest.

The genera *Populus* and *Platanus* are not found in the living forest, although *Populus dimorpha* occurs along the west coast of Mexico approximately as far south as latitude 25°N. The complete absence of *Platanus* from the flora of New Galicia (see pg. 34 below) is one of those peculiar quirks of plant distribution. *P. racemosa* occurs along the stream courses south of Nogales to about latitude 30°N, while *P. mexicana* and *P. lindeniana* are common in the tropical forests east of the summits of the Sierra Madre Oriental. The well-preserved pollen flora contains such tropical genera as cf. *Bombax*, cf. *Manilkara,* and cf. *Engelhardtia*.

In the winter-dry warm temperate to tropical flora of southern Asia, Indo-China to India, are found:

Aleurites moluccana
Asplenium sp.
Castanea seguinii
Eugenia (Jambos) jambos
Glyptostrobus heterophyllus
Ilex rotunda

Machilus (Persea) velutina
Derris trifoliata
Schefflera heterophylla
Sterculia javanica
Lygodium palmatum
Thelypteris sp.

Thus it appears that many of the correlative living species are characteristic of tropical, winter-dry climates, either of southeastern Asia or southwest Mexico.

In the warm temperate or temperate forests of southeastern Asia and Japan, and the southern Appalachians are found:

Castanea spp.	Derris sp.
Diospyros virginiana, kaki	Liquidambar styraciflua, formosana
Dipteronia sinensis	Prunus serotina, campanulata
Ilex rotunda	Symplocos tinctoria, glauca
Juglans nigra, cathayensis	Zelkova serrata

Tropical deciduous or semi-deciduous forests do not seem to be of importance in the vegetation of southern China. The temperature limits in southern China corresponding to those of the tropical, semi-deciduous forests of southwest Mexico are occupied by the warmer phases of the "evergreen, broad-leaved forest" of Wang (1961:143). This forest is dominated by numerous species of evergreen oaks, Schima, and laurels. Aycross genera found in this forest are: *Castanopsis, Engelhardtia, Eugenia, Dendropanax, Ilex, Liquidambar, Machilus (Persea), Saurauia* and *Symplocos. Diospyros, Dipteronia, Toona (Cedrela), Populus, Salix* and *Zelkova* occur in the broad-leaved, deciduous forest to the north, which, in its abundance of oaks, maples and linden resembles the forest of the middle and southern Appalachians.

Farther to the west and southwest in India and Burma, tropical deciduous and semi-deciduous forests are common, and are sometimes designated as "monsoon forests." The typical semi-deciduous tropical forests are dominated by Dipterocarpaceae, *Tectona* and *Terminalia*. Aycross genera found in these forests are: *Bombax, Diospyrôs, Eugenia, Ilex, Schefflera, Sterculia*, and *Toona (Cedrela)*. (Haden-Guest, Wright, Teclaff, 1956:461–468.) This area experiences a pronounced dry season in contrast to southern China which receives winter rain from traveling cyclonic storms.

CLIMATE AND VEGETATION OF THE SEMI-DECIDUOUS FORESTS OF SOUTHWEST MEXICO

As an approximate representation of the vegetation and climate of the western Wind River Basin 48 million years ago, descriptions of the vegetation and climate of the semi-deciduous forest of southwest Mexico are given.

Rzedowski and McVaugh (1966) in their analysis of the vegetation of Nueva Galacia in southwestern Mexico have defined about 14 vegetation types. Pennington and Sarukhan in their study of the tropical trees of Mexico (1968) have developed a slightly different classification of the arboreal vegetation (pgs. 4–44) although the two classifications are in essential agreement. The term Nueva Galicia is derived from the old Spanish administrative state of that name, which included all of Nayarit, Jalisco, Colima and Aguascalientes, with the north half (approximately) of Michoacan and small portions of other adjacent states. It was bounded on the north by the latitude of 22°30', on the south by 18°30', on the west by the Pacific Ocean, and on the east by a line drawn approximately through Leon and Apatzigan. The vegetation in this region is complex, ranging from coastal mangrove swamps through lowland tropical vegetation to montane pine-oak, montane fir-alder and high altitude grasslands and shrub, up to altitudes of 4000 meters.

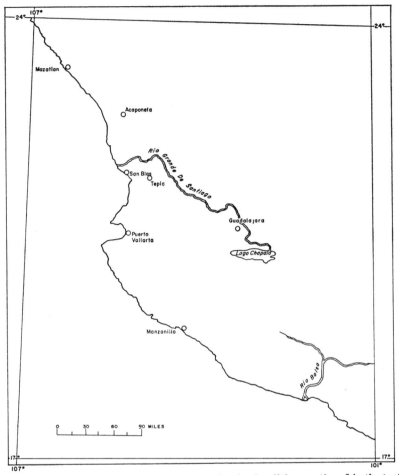

Fig. 5. Sketch map of southwestern Mexico showing localities mentioned in the text.

Certain aspects of the "Bosque Tropical Subdeciduo" (1968:15) are pertinent to the paleoecology of the fossil flora. The authors describe this vegetation as follows: "Among the types of vegetation which are described in the present contribution this is indubitably the most luxuriant [and] the most complex for its structure as well as for its floristic composition. Its physiognomy and phenology place this formation in a situation intermediate between the bosque tropical perennifolia and the bosque tropical deciduo. Although the great majority of species lose their leaves during the dry period there are many trees which do not defoliate completely, and others which are deciduous for only a short period, at times only a few weeks. The height of the dominant tree stratum is invariably greater than in the case of the bosque tropical deciduo, and equal to that in the abundance of lianas, epiphytes, and shade plants." This vegetation type may be seen in the coastal strip west of the western Sierra Madre between San Blas and Puerto Vallarta, at altitudes from 0 to 1200 meters.

In describing the climate of "Nueva Galicia" Rzedowski and McVaugh say:

The varied topography of New Galicia has its reflection in a notable diversity of climates. The equatorial temperature touches the extreme south of the area, and, on the other hand, in the peaks of Tancitaro and the Colima volcano the limit of arboreal vegetation is reached. Truly humid or truly arid climates are lacking but there exists all the range of intermediate climates. As notable climatic characters the absence of marked temperature seasons should be emphasized [as also] the presence of two well defined moisture seasons.

The temperature appears to reach its highest values in the valley of the Tepalcatepic River, part of the low area along the Balsas, where the average annual temperature registers as much as 29°C. Along the coastal strip the heat is less intense and the annual average temperature remains between 25°C and 27°C. At neighboring altitudes of 1600 meters [5248 feet] the values are around 20°C [Guadalajara, 19]). In agreement with Gutierrez Vasquez (1959, p. 57) the average temperature gradient below 1200 meters [3036 feet] is 0.0031 [degrees Centigrade per meter], and above 1200 meters this value raises to 0.0045°C per meter.

The zone free from freezing is, in general, below 1200 meters altitude but, in some localities ascends to 1600 meters above the level of the sea.

The warmest month is commonly May or June, and January is the coldest; the differences between the average temperatures varies between 2.5°C and 10°C, increasing in general from the coast toward the interior. The maximum temperatures reach values around 50°C in the warmest areas ... the area receiving the least precipitation is localized toward the extreme northeast where the average annual rainfall is less than 500 mm. (385 mm. is the lowest average recorded). The most humid strip is in some of the mountains near the coast in Jalisco and Nayarit with precipitation of more than 1500 mm. (there are areas in which the precipitation is more than 2000 mm. but these must be very limited). In general values greater than 1000 mm. are restricted to the mountainous regions, on the other hand the zones where the rainfall is between 750 mm. and 1000 mm. occupy the major area of New Galicia. The lower parts of the depression corresponding to the basin of the Balsas and the small coastal plains of Tomatlan and of Tecoman, with precipitation less than 750 mm. are also markedly dry, thanks to the elevated temperatures which prevail there ... in general more than 90% of the rain falls, on the average, in the months May to October, and thus the year is divided into a rainy and a dry period; the duration of the last varying from 5 to 8 months. The rainfall is of the torrential type of short duration, and occurs usually in the afternoon. [The translation is mine as are any errors occurring therefrom.]

A large proportion of the species in the tropical, semi-deciduous forest of southwest Mexico are deciduous for longer or shorter periods during the dry season from November to May. A survey of this forest type (Rzedowski and McVaugh, 1966:16–22) shows 59 percent entire (41 percent non-entire). This is about the same as the low altitude Taiwan forest (Wolfe, 1971:34) and 14 percent greater than the fossil flora. The semi-deciduous forest of southwest Mexico grows in a climate which is clearly tropical and not subtropical; the average annual temperatures of some stations in the area are: San Blas, 24°C, altitude 9 m.; Ahuacatlan, 22°C, altitude 1003 m.; Tepic, 21°C, altitude 922 m.; Guadalajara, 19°C, altitude 1589 m. Light frosts have occurred in the Guadalajara area but the other stations are frostless. (See Mora and Jáuregui, 1965.)

The Mesophyll montane forest, bosque mesofila de montaña (Rzedowski and McVaugh, 1966:69–72), grows in southwest Mexico between altitudes of 800 and 2400 meters. It occupies about the same altitudinal range as the pine-oak forest, but is typical of protected canyon sides where the humidity is higher than on the exposed ridges occupied by the pine-oak vegetation. It is best developed between altitudes of 3500 and 4500 feet on the mountain sides facing the Pacific. An excellent example of this forest may be seen along the road toward Jalcocotán, 7 miles southwest of Tepic in the canyons leading west from the volcanic mountains. The montane forest interfingers with the semi-deciduous forest below. It contains many characteristically "temperate" genera: *Carpinus, Clethra, Fraxinus, Ilex, Juglans, Meliosma, Ostrya, Prunus, Quercus, Salix, Tilia, Symplocos, Rhus, Parthenocissus, Vitis*, among others. A survey of 96 species showed 60 percent non-entire margins (40 percent entire), and 55 percent deciduous. The lower part of this forest is frost-free. There are many genera which may be considered as tropical in aspect: *Dendropanax, Phoebe, Trophis, Calyptranthes, Eugenia, Conostegia, Inga, Piper, Siparuna, Triumfetta*, for example. A marker for the lower montane forest is the presence of three species of *Oreopanax* which are relatively abundant, and conspicuous due to their large, palmately dissected leaves. Several of the more temperate-appearing genera of the fossil flora occur in this forest, but I have seen no *Populus*, and *Platanus* is entirely absent.

The fossil flora with 45 percent entire margined leaves and 61 percent deciduous compares with 40 percent entire margins and 55 percent deciduous of the bosque mesofila de montaña, and 59 percent entire, 60 percent deciduous for the semi-deciduous forest. Thus the fossil flora shows about the same proportion of deciduous species as the tropical, semi-deciduous forest, but the percentage of entire margins is closer to that of the lower montane forest indicating a somewhat lower average annual temperature than that of the tropical semi-deciduous.

CLIMATIC INDICATIONS OF THE FOSSIL FLORA

The most satisfactory climatic classifications for the use of the paleobotanist are probably those founded on or derived from the Köppen system (W. Köppen, 1931: Köppen-Geiger, 1954). Such systems as those of Thornthwaite (1948) are not convenient for the paleobotanist, since they require data not available for the study of Tertiary floras from which to derive such concepts as moisture index and thermal efficiency. It is a temptation for the paleobotanist to over-interpret his data. Quantitative values for Paleogene climates, in terms of our present knowledge, can be derived in only the most general way from the data of fossil plants. The general temperature regime under which an early Tertiary flora lived can probably be derived with fair accuracy from a study of the living correlatives of the fossil species. Estimates of moisture conditions must be more approximate since these depend so much on relative amounts of cloudiness, wind movement, type of precipitation, and seasonal distribution of precipitation.

The terms tropical and subtropical have been so loosely used that definitions are always in order. (1) Tropical, mean annual temperature above 22°C, mean temperature of the coldest month above 18°C; (2) Paratropical, mean annual temperature above 22°C, mean temperature of the coldest month 15°C to 18°C; (3) Subtropical, mean annual temperature 15°C to 19°C, mean temperature of the coldest month 6°C to 10°C; (4) Warm temperate, mean annual temperature 11°C to 16°C, mean temperature of the coldest month, –3°C to 5°C (Wolfe, 1969: 39–44). Wolfe has coined the term paratropical (near tropical) in order to describe a climate under which some of the forests of the Eocene to Middle Oligocene appear to have existed (fig. 6). These are typified by the Lower Ravenian in southwest Alaska, Steel's Crossing near Seattle, Washington, and Bilyeu Creek in western Oregon. Mai's climatic determinations for the "mastixoid" flora of middle Europe during the Miocene are as follows (1964:148): average temperature, 20°C to 23°C, average January temperature, 10° to 18°C, average July

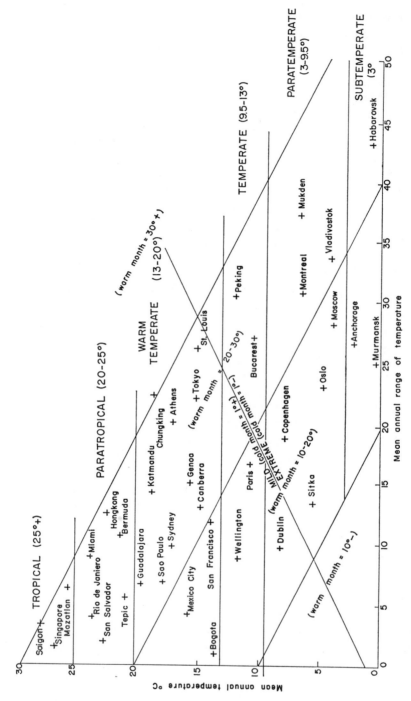

Fig. 6. Tentative classification of climates based on mean annual temperature and mean annual range of temperature. Modified slightly after Wolfe.

temperature 25° to 30°C. The precipitation was estimated as: annual, 130 to 200 cm.; January 5 to 10 cm.; July, 20 to 30 cm. These temperature estimates place the Mastixoid floras as paratropical according to the ranges given above. The term tropical should not be used for climates where frosts occur; these climates are frostless. Frosts occur in subtropical climates but they are not severe or of long duration. In warm temperate climates freezing temperatures are probable during several months of the year. The climate of central Alabama would be considered warm temperate on the basis of low winter temperatures even though the annual mean is 18°C. The climate of southwest Mexico between Manzanillo and Mazatlan up to an altitude of 3000 feet is essentially tropical under the definition above. From 3000 to 4500 feet altitude the climate in this area is, for the most part, paratropical.

Kira (1948, 1949) has developed climatic measures which he calls Warmth Index and Coldness Index, based on the idea of "total effective temperature" (see Tanai, 1970:430–431). The warmth index is obtained by taking the sum of the differences of the mean monthly temperatures above 5°C for the year. Said in another way it is the sum of each mean monthly temperature minus 5°C. The coldness index is the sum of the mean monthly temperature below 5°C. The dividing temperature of 5°C is well chosen since, in general, the growth of higher plants ceases or is markedly slow below this temperature, and above it plant growth tends to proceed more rapidly. Tanai says, "The two indices play an important role for limited distribution of modern forests; WI represents a positive action for plant life while CI represents a negative one" (1970:430). I have calculated the warmth index for three stations in southwest Mexico: San Blas, elevation 10 feet, 241.5°; Tepic, elevation 3100 feet, 193.8°; Guadalajara, elevation 5200 feet, 167°. The unit in which these figures are given is the degree Centigrade. The WI of 190° gives the approximate northern boundary of what Wolfe (1969:41, 43) has called the paratropical rain forest in southeast Asia, and this index is also a fairly satisfactory upper altitudinal limit for the tropical semi-deciduous forest of Nayarit, Mexico.

Thus the fossil flora appears to have the closest correspondence with the transition forest between the tropical semi-deciduous and the mesophyll montane forest of southwest Mexico. The climate of Tepic, altitude 3100 feet (950 m.), average annual temperature, 22°C (61.6°F) is probably the closest correlative to that of the northwestern Wind River Basin in the early Middle Eocene, as close as can be found in the present climates of the northern hemisphere.

Again, in terms of floristic composition, the average annual precipitation could hardly have been less than 35 inches, nor more than 55 inches. Similar vegetation flourishes today with that amount of summer rainfall (Mazatlan, 33 inches; San Blas, 58 inches; Tepic, 54 inches). This estimate of from 35 to 55 inches contrasts with the present annual precipitation of 9 inches at Dubois, Wyoming.

I know of no reliable method for accurately estimating the average annual temperature range in the Wind River Basin during the Eocene. The region at present has an extreme continental type of climate; average annual range 40°F at Dubois, extremes 98°F and –48°F. The Eocene mean annual range may have been as much as 15°C, but it was probably less, closer to 5° or 10°C.

Paleobotanists, in the past, may have been overcautious in the use of the term tropical for the indications of Cretaceous and Paleogene floras. The Kisinger Lakes-Tipperary flora, if found growing today would be called tropical without hesitation. This fossil flora, as well as some of the early Oligocene floras of southwest Alaska, illustrate the extraordinary contrast between the vegetation of the Eocene and that of the present in extra-tropical latitudes.

On first consideration it may seem strange that the early Eocene Chalk Bluffs flora from the central Sierra Nevada shows a rather marked resemblance to the existing flora of the *eastern* Sierra Madre of Mexico between the latitudes of 19° and 23°N; while the early Middle Eocene flora of the Wind River basin, about 750 miles to the northeast of California, should have a strong resemblance to the living flora of the *western* Sierra Madre at the same latitude (19° to 23°N). This can be explained on the basis of the indicated climatic changes of the early Tertiary. The known Early Eocene floras of North America from the Gulf coast to northwestern Washington, indicate a humid climate with annual rainfalls of approximately 50 inches, well distributed through the year. The Wind River flora of late Eocene age, 4 to 5 million years older than the Kisinger Lakes flora and in the underlying sediments, is a generalized flora of warm temperate aspect with little or no indications of a dry season. The Kisinger Lakes flora, with its more tropical aspect and implications of a marked dry season, shows the beginnings of vegetational change in response to the climatic change in the direction of dryer and warmer climates in the central Rockies. The late Eocene pollen floras of the area and the megaflora from Elko, Nevada, show clear evidence of a rather pronounced dry season. Thus the difference in the climatic indications of the Chalk Bluffs flora with respect to the Kisinger Lakes flora is paralleled by the existing climatic difference between Tepic and Orizaba, Mexico.

The uplifts of the Early Eocene had been subjected to several million years of erosion by the Middle Eocene, and thus the ranges bordering the Wind River basin must have been reduced to very moderate altitudes at the time the flora was deposited. The elevation of the basin surface seems not to have exceeded 1000 feet. The characters of the sediments and certain aspects of the flora indicate that topographic contrasts were of moderate degree as contrasted to the present rugged relief.

For emphasis it may be repeated that the species lists show nearly half of the total correlative living species occurring in the bosque tropical subdeciduo of southwest Mexico and in the warmer aspects of the bosque mesofila de montaña. The tropical subdeciduous forest grades upward into, and interfingers with, the montane mesophyll forest in discontinuous areas, mainly along the more protected and more humid canyons and barrancas which the streams have eroded along the western slopes of the Sierra Madre Occidental.

Thus the climate indicated by fossil flora may be designated as between paratropical and subtropical, meaning more tropical than warm temperate but not truly lowland tropical. The present climate of the Pacific side of the Mexican Sierra Madre Occidental between the latitudes of 23°N and 19°N, and at altitudes between 2500 and 4000 feet, would furnish a favorable setting for the entire group of correlative living species. The conclusion in the above discussion are summarized below.

1. Habitat: floodplain of a moderately large stream of low gradient, which traversed the area from the WNW between westerly trending ranges of hills on the north and south, the elevations probably nowhere higher than 2000 feet above the valley floor.
2. Elevation of the basin floor not exceeding 1000 feet above sea level.
3. Volcanic centers not far to the north and northwest, some of them extremely active at times; the eruptions partly of the central vent, explosive type and partly from fissures.
4. Average annual temperature between 19°C and 23°C; average temperature of the coldest month not below 15°C. Paratropical to subtropical.
5. Little or no frost.
6. Warmth Index near 190°.
7. Warm summers, with the temperature of the warmest month as high as 27°C.
8. Average annual precipitation not below 35 inches (875 mm.) nor above 55 inches (1125 mm.).
9. Seasonal precipitation, with dry winters and rainfall in the summer season.
10. Vegetation: extra-tropical or paratropical, semi-deciduous forest.
11. Low elevations around the basin floor occupied by a mixed, coniferous forest with pine, oak, Bombacaceae, various Leguminosae, and Juglandaceae (*Carya, Engelhardtia*).
12. pH of the substratum near 7 (6.5 to 7.5) caused by the abundance of silica in the groundwater produced by weathering of volcaniclastics.

The causes of the climatic change, the difference in climate between the Middle Eocene and the present, appear to be extremely complex. These causes probably derive from some combination of the following:

(1) increased elevation of the continents and increase in continental areas;
(2) possible changes in the inclination of the earth's axis;
(3) changes in latitude and longitude due to continental drift;
(4) changes in the motion of the earth with respect to the sun;
(5) possible changes in solar radiation.

Number (1) can actually be measured with some degree of accuracy and its climatic effects are rather clear. Effects of the other four changes on earth climates can be expressed in only the most general way, and must remain, in part at least, in the realm of hypotheses.

RELATIONS TO OTHER FOSSIL FLORAS

Before discussing the relations to other Eocene floras in the area a statement concerning the age of the flora is in order. Obradovich has obtained a satisfactory K/Ar date of 46.2 m.y. from the base of the Wiggins Formation and a date from the White Pass bentonite layer of 49.3 m.y. This places the age of the Kisinger Lakes floral zone as approximately 48.5 m.y. or basal Middle Eocene (Bridger). (See fig. 3.) The Tipperary locality is from 50 to 75 feet above the base of the contact between the Aycross and Wind River formations. The two Middle Eocene localities can hardly differ in age by any significant amount. The differences in the two collections can be explained on the basis of local floristic differences. They are essentially the same flora. If there is any difference in age the Tipperary flora, on the basis of the fossils, might be slightly older. Therefore the flora is early Bridger in age (Evernden, Savage, Curtis and James, 1964:167).

There are two fossil floras in the central Rockies which show significant resemblances to the Kisinger Lakes-Tipperary flora: the Wind River and the Green River. The flora under discussion is intermediate between the two in its general floristic and vegetational composition. It may be well to note here that a list of similar or identical species can reveal only a partial account of the relationships between two fossil floras (table 2). The more important and critical aspects of the floras are given by the ecologic indications and floristic composition of the whole floras. The Wind River flora is a mesic, warm-temperate to subtropical flora indicating a greater rainfall, and a somewhat lower average annual temperature than the Kisinger Lakes-Tipperary flora. The Wind River flora is of Lost Cabin age (Keefer, 1957:190–191) or late Early Eocene. In terms of the K/Ar dating it is from 50.2 to 52 m.y. old.

In the same general area, that is the northwestern Wind River basin and environs, there are 4 localities for the Lost Cabinian flora: (1) the original Wind River locality from 10.5 to 13 miles west of Dubois on the north banks of the Wind River, Sections 3, 4 and 10, T 42 N, R 108 W, Warm Spring Mountain and Esmond Park Quadrangles; (2) the Coyote Creek locality, about 16 miles west of Dubois on the north slope of the Wind River Range, Section 10, T 42 N, R 108 W, Fish Lake Quadrangle; (3) the Fish Lake locality in the same general area, Section 6, T 41 N, R 109 W, Fish Lake Quadrangle; (4) the Three Tarns locality, about 22 miles west of Dubois in the western end of the Wind River Range, Section 10, T 42 N, R 108 W, Sheridan Pass Quadrangle. These floral horizons do not differ much in age, the Coyote Creek locality appears to be the youngest and is approximately one-half million years younger than the original Wind River locality, on the basis of the stratigraphy.

These Lost Cabinian floras (Wind River) although only 2 to 4 million years older than the Kisinger Lakes flora and in the same general area are considerably different in composition and ecologic aspect. There is a marked relationship to the living floras of warm temperate southeastern Asia, shown by such genera as *Actinidia, Eustigma, Firmiana, Magnolia, Paliurus, Pterocarya, Platycarya* and abundant laurels. The early Eocene forest was much like the transition vegetation between the mixed mesophytic and evergreen broad-leafed forests of Wang (1951: 95–164). There is no definite evidence of a pronounced dry season. The pH of the substratum must have had a lower value than that of the Kisinger Lakes flora in which the absence or rarity of such mega-fossil genera as *Magnolia, Nyssa, Tilia* and Ericaceae suggests a pH of 6.5 or possibly more, probably controlled by the abundance of volcaniclastic material. Many of the characteristic Kisinger Lake species have, so far, not been found in the Lost Cabinian floras: *Aleurites, Acrostichum, Canavalia, Eugenia, Luehea, Platanus brownii,* and *Sterculia* for example. The only Platanoid leaf in the Lost Cabinian floras is the typical "*Platanophyllum*"; there is no trace of the narrow-lobed, serrate forms that are abundant in the Kisinger Lakes and Green River floras. Some of the Kisinger Lakes species show clear evolutionary changes from the Wind River forms. This is illustrated by *Dendropanax, Populus wyomingiana* and *Symplocos.* The Kisinger Lakes flora is a derivative of the older, underlying Lost Cabinian floras; the changes are the result of speciation, together with dispersal and probably some extinctions, caused by a change to warmer and dryer conditions.

Similarly the Green River flora (MacGinitie, 1969) is a derivative of the Kisinger Lakes-Tipperary flora showing a progressive adaptation to subhumid conditions. A K/Ar date of 45 million years has recently been determined in the Green

River Formation of northern Colorado several hundred feet above the mahogany ledge (personal communication, W. Culbertson, April 1971). The Green River locality at Wardell Ranch would be only slightly older than this, not more than 0.5 m.y. older (MacGinitie, 1969:81). Several of the Green River species are identical or closely similar to characteristic Kisinger-Tipperary species: *Aleurites, Acrostichum, Eugenia, Luehea, Populus* cf. *wyomingiana, Salix* and *Zelkova* for example. *Platanus wyomingiana* and *P. brownii* are closely similar or perhaps identical species. The Kisinger-Tipperary flora lacks such typical Green River forms as *Allophylus, Bursera, Cardiospermum,* evergreen oaks, *Mahonia, Mimosites, Rhus* and *Styrax,* which appear to have flourished in the less humid climate to the south. The numerical dominants of the Green River flora: *Mimosites, Zelkova, Rhus, Platanus, Cardiospermum, Allophylus, Populus, Leguminosites, Sapindus* and *Salix,* give this flora a much different aspect from that of the Wind River Basin. The Paleogene of the central Rockies was a time of comparatively rapid changes in floras and climates.

The Little Mountain flora in the Currant Creek Ranch Quadrangle, Township 13 North, Range 106 West, Section 27 (MacGinitie, 1969:62, 66–67, no. 1, fig. 1) has recently been dated by K/Ar values at 49 m.y. (Culbertson, personal communication, May 1971). This is a typical Green River type flora with strong subhumid indications. It is of nearly the same age as the Kisinger Lakes-Tipperary flora but considerably different in aspect. This raises the question of how two such different floras could exist at about the same time in the same general area. The cause must have been some topographic-climatic barrier. This could not have been the Uinta Range since similar floras, the Little Mountain and the Green River, existed on both sides of it. The best explanation seems to be that an elevated area existed to the west in the Great Basin geanticlinal area (MacGinitie, 1969:62–63) which cut off the ingress of moist Pacific air from about latitude 43°N south as far as latitude 34°N or even farther. Such a rise would cause descending air currents on the east side giving rise to warmer air of lower relative humidity. The present Sierra Nevada and perhaps Wasatch Ranges were apparently of little or no effect as climatic barriers at the time. During the Early Eocene and early Middle Eocene the Wind River basin was much moister than the area to the south. Studies by Bradley of the paleohydrology and paleoclimatology of the Green River Formation have revealed that in a period of about one million years during the Middle Eocene, the climate of [southern] Wyoming changed from moist to arid (as arid as the Great Salt Lake today) and then became moist again (Bradley, 1961:25; Bradley and Eugster, 1969:21–25).

Columns 4, 5 and 6 (table 3) are given to show briefly the occurrence of early Middle Eocene floras elsewhere in the area. These sites have not as yet been extensively collected. The florules from the Fort Hill Quadrangle locality (Section 25, T 24 N, R 114 W) and the Rhodes Ranch locality (Cottonwood Creek) indicate that they are a part of the same floral horizon as the Kisinger Lakes-Tipperary. They contain *Platanus* cf. *brownii, Acrostichum, Luehea, Lygodium, Aleurites, Salix* cf. *molesta,* and *Lygodium.* The Valley flora near the village of Valley on the South Fork of the Shoshone River, contains some typical early Middle Eocene species but also a few species somewhat closer to the Wind River flora, indicating

	1	2	3	4	5	6	7	8	9
ACALYPHA CIRQUENSIS	X				?				
ACROSTICHUM HESPERIUM		X		X	X		X		X
ALEURITES FREMONTENSIS	X	X		X		X	S		
AMPELOPSIS TERTIARIA		X							
APEIBA IMPROVISA	X	X							
ARISTOLOCHIA SOLITARIA	X								
ASPLENIUM EOLIGNITUM	X	X	S				S		X
CANAVALIA DIUTURNA	X								
CARPITES ARALIODES	X	X							
CASTANEOIDES AEQUALITA	X		X						S
CEDRELA SCHIMPERI	X	X			X	X	S		S
CHAMAECYPARIS sp.	X		X						
cf. CISSUS sp.	X								
CORNUS sp.	X		X						
DENDROPANAX LATENS		X	S	?					
DIOSPYROS MIRA	X								
DIPTERONIA WYOMINGENSE	X	X					S		
EQUISETUM TIPPERARENSE	X	X			X	X	S		
EUGENIA AMERICANA	X						X		S
GEONOMITES HAYDENII			X						
GLYPTOSTROBUS EUROPAEUS	X								
ILEX SCLERA	X								
JUGLANS ALKALINA			X						
LAUROPHYLLUM FREMONTENSIS	X	X							
LAUROPHYLLUM QUOTIDIANA	X		S					X	S
LEGUMINOSITES INLUSTRIS	X								
LEGUMINOSITES MIRA	X								
LEGUMINOSITES OCCULTA	X								
LEGUMINOSITES WYOMINGENSIS	X								
LIQUIDAMBAR CALLARCHE		X	X				X		S
LUEHEA NEWBERRYANA	X	X		X			X		
LYGODIUM KAULFUSSII	X	X	S		X	X			X
PLATANUS BROWNII	X	X			X	X	S	X	
PLATANUS INTERMEDIA	X	X							
POPULUS QUINTAVENA	X								
POPULUS WYOMINGIANA	X	X	S	X	X		S		S
PROTEACIPHYLLUM MINUTA	X								
PRUNUS NOTA	X		S				S		
SABALITES FLORISSANTI	X	X					X		
SABALITES POWELLII	X	X							
SALIX MOLESTA	X	X		X	X		S		
SALVINIA PREAURICULATA		X	X						X
SAURAUIA PROPIA	X								
SCHEFFLERA INSOLITA	X								S
SERJANIA RARA	X								
SPARGANIUM ANTIQUUM	X	X	X				X		
SPIRODELLA MAGNA	X					X			
STERCULIA SUBTILIS	X								
SYMPLOCOS INCONDITA	X	X	S				S		
THELYPTERIS IDDINGSII	X	X			X				
THELYPTERIS WEEDII	X	X							X
ZELKOVA NERVOSA	X	X	S	X			X	X	
IDENTICAL SPECIES	44	35	9	6	8	6	7	4	5
SIMILAR SPECIES			8				10		7

Table 3. Distribution of Kisinger Lakes species and genera in other fossil floras. 1. Kisinger Lakes. 2. Tipperary. 3. Wind River. 4. Rhodes Ranch. 5. Valley. 6. Fort Hill. 7. Green River of northern Colorado. 8. Bear Paw Mountains. 9. Claiborne-Puryear.

Tentative correlations of some North American Tertiary floras.

K/Ar Age	Pacific Coast	Gulf Coast	Northwest Wyoming	Nevada-Idaho-Montana	Colorado-Nebraska southern Wyoming
3					
12					
	Trout Creek Sucker Creek Callowash			Buffalo Canyon-Fingerrock	Kilgore
26					Creede
	Crooked River-Bridge Creek Willamette Goshen La Porte	Vicksburg		Ruby Metzel Ranch	
38					Florissant
	Lower Cedarville ?Clarno nut bed Chalk Bluffs	Yazoo-Jackson Claiborne Wilcox	Kisinger Lakes Early basic breccia Wind River Early Acid Breccia	Copper Basin	Green River Little Mountain Golden Valley
56					
		Wilcox Midway	Fort Union		Denver
65					

TABLE 4.

that it is transitional between the Kisinger Lakes and the Wind River. There are several localities in the Bear Paw Mountains of northern central Montana which also contain typical Kisinger Lakes species. These floras will be described more fully later.

The Eocene fossil flora of Yellowstone Park was described by Knowlton (1899), and, after the custom of the day, the plant names were more stratigraphic than botanic. Dorf (1960:258–259) has given a revised list, from which the following species names were taken, showing similarities among the floras. A thorough revision of the Yellowstone floras is needed.

AYCROSS FLORA	EARLY BASIC BRECCIA	EARLY ACID BRECCIA
Lygodium kaulfussii	Lygodium kaulfussii	Lygodium kaulfussii
Asplenium eolignitum	Asplenium iddingsii	Asplenium iddingsii
Thelypteris weedii	Dryopteris weedii	Dryopteris weedii
Monocot spp.	Musophyllum complicatum	Musophyllum complicatum
Aleurites fremontensis	Aralia notata	Aralia notata
Acalypha cirquensis	-----	Acalypha aequalis
Castaneoides aequalita	Castanopsis longipetiolatum	Castanopsis longipetiolata
Cornus sp.	Cornus wrightii	Cornus kelloggii
Apeiba improvisa	Ficus mississippiensis	Ficus pseudopopulus
Diospyros mira	Diospyros lamarensis	Diospyros lamarensis
Cedrela schimperi	Juglans schimperi	Juglans schimperi
Laurophyllum quotidiana	Persea pseudocarolinensis	Persea pseudocarolinensis
Platanus brownii	Platanophyllum angustiloba	Platanophyllum angustiloba
Platanus intermedia	Platanus appendiculata	Platanus appendiculata
Salix molesta	Salix cockerellii	Salix cockerell

Considering their contiguity (about 80 miles separation) and their similarity in age, the lack of close similarity in the floras is rather surprising. This may be in part because of the lack of intensive herbarium comparisons in the study of the Yellowstone flora. The use of names by Knowlton and Lesquereux leaves the real identity of the fossil in doubt. However it seems probable that both the Yellowstone floras are older than previously estimated. The Chalk Bluffs fossil flora of California (MacGinitie, 1941) I now believe is of early Eocene age and probably as old as Lysitian (MacGinitie, 1969:127). The presence of such species as *Magnolia spectabilis, Phytocrene sordida*, "*Platanophyllum*" *whitneyi, Platanus montana, Artocarpus lessigiana, Gordonia egregia*, "*Laurophyllum*" *litseaefolia*, and *Thouinopsis myricaefolia* indicate that both the Yellowstone floras as now known are early Eocene in age, with the flora of the early acid breccia being about the same age as the Chalk Bluffs flora. The flora of the early basic breccia is slightly younger than the flora of the early acid breccia and somewhat older than the Aycross flora. It also seems probable that the Yellowstone floras were deposited at a somewhat higher elevation than the Aycross flora, and this may account for some floristic differences.

The Middle Eocene (Claiborne) flora of the Gulf Coast is of nearly the same age as the flora from the Wind River Basin. Berry (1930a) in his revision of the "Lower Eocene Wilcox flora" gives species lists of a large number of fossils recovered from the Puryear clay pit in western Tennessee and from the Holly Springs sand. This horizon is now considered to be of Middle Eocene age. Considering the distance between the two floras and the long separation during the Cretaceous, a strong resemblance should not be expected. In addition to this the Tennessee flora is a tropical strand flora dominated by legumes, laurels, soapberry and "Dryophyllum," while the Wyoming flora occupied the floodplain of a moderately large river, in a hilly, interior region, and was dominated by *Cedrela, Populus, Platanus* and *Lygodium*. In dealing with Berry's published accounts as a source of information concerning the relationships of the two fossil floras serious difficulties arise: (1) hasty and erroneous identifications, (2) over-speciation, (3) inadequate descriptions and illustrations, (4) confusion regarding correct stratigraphic placement. Also a number of the types have disappeared. With regard to identifications the illustrations of *Hicoria crescentia*, plate 156 (Berry, 1930a) and *Grewiopsis tennesseensis* (pl. 15) serve to illustrate this type of difficulty. As illustrations of over-speciation Berry lists 14 *Cassia*, 6 *Cedrela*, 6 *Dryophyllum*, 25 *Ficus*, and 10 *Sapindus*. The Claiborne flora must necessarily be completely reworked. It is finely preserved, both mega- and microflora, even as far as excellent preservation of cuticle. Dilcher, at Indiana University, has made good progress with identifications of cuticular preparations. The two fossil floras appear to possess these genera in common: *Acrostichum, Asplenium, Aristolochia, Cedrela, Dendropanax, Diospyros, Eugenia, Glyptostrobus, Ilex*, two legumes, *Liquidambar, Lygodium, Persea, Platanus, Sabalites, Salvinia, Schefflera, Sterculia*, and *Thelypteris*. Some of these genera are wide-ranging in space and time, but specific identities cannot now be established. About all that can be said is that there is a climatic resemblance, and also that the Claiborne flora shows just the type of strand flora to be expected contemporaneously with the inland Kisinger Lakes-Tipperary flora.

There is a group of genera which seem to be of widespread occurrence in the northern hemisphere from the Eocene to Middle Oligocene. These are illustrated by the genera common to the Wyoming flora and the flora of the Ube coalfield in Honshu, Japan (Huzioka and Takahasi, 1970). The Honshu fossil flora contains 63 species of woody dicots of which 60 percent were deciduous and 65.4 percent entire. The vegetation type can be variously designated as a "laurisilva"; oak-laurel forest; or overgreen, broadleafed sclerophyllous forest. Its general aspect then would have been considerably different from the Wyoming flora. The genera common to the two fossil floras are:

Acrostichum	Ilex
Castanopsis	Liquidambar
Cedrela	Machilus-Persea
Dendropanax	Platanus
Diospyros	Sabal
Glyptostrobus	Zelkova

No *Populus* or *Salix* has been reported from this flora. Among pollen and spore genera, commonly found in the Northern Hemisphere during the later Paleogene are *Carya, Engelhardtia, Glyptostrobus, Juglans, Lygodium*, Palmae and *Pterocarya*. The genera *Castanopsis, Cinnamomum, Tetrastigma* and *Symplocos* appear to have been widespread at temperate latitudes in the fruit-seed and leaf floras of the later Paleogene and early Neogene European, and Asian floras.

The European megafloras of Eocene age, with the exception of the fruit and seed floras of the London Clay and the Braunkohl, are much in need of revision; no European Tertiary flora similar to the Wyoming flora is known to the writer. The

Russian paleobotanists have produced a large amount of research and Pokrovskaia (1966:277–282) gives lists of pollen species from southwestern U.S.S.R. Among the forms listed from the Middle Eocene are *Glyptostrobus*, Palmae, *Carya, Pterocarya, Engelhardtia, Castanea*, Ulmaceae (*Zelkova* ?), Lauraceae, *Ilex*, Tiliaceae, Myrtaceae, Araliaceae. The floras are of subtropical or warm temperate aspect. There is no evidence of seasonal precipitation nor any close resemblance to the Middle Eocene flora of the Wind River Basin. Berry (1924:37) believed that he saw a distinct likeness to the Claiborne flora in the Upper Lutetian flora of Novale in Venice, but, owing to the early date (1854) of the paper the resemblance can be described as only general. Penny has mentioned a "phase in palynologic history marked by considerable preoccupation with nomenclative experiment. In conse-quence much of the accumulated literature [European] is virtually a cemetery of abandoned taxonomic fashions, invalid types, quasi-trinomials and involved new combinations" (in Tschudy and Scott, 1969:357). Thus, for the present, because of linguistic barriers, lack of modernized paleobotanical research in the earlier Paleogene megafloras of western Europe, and inadequate taxonomy of pollen species, significant comparisons between the Wyoming and European floras must await further reseach.

[The following chapter is the work of Estella Leopold, palynologist of the United States Geological Survey at Denver, Colorado. It is one of the products of Dr. Leopold's extensive research on and experience with the pollen floras of the Rocky Mountain region. Our collaboration in the study of Tertiary floras in the central Rockies extends over a period of a dozen years. (Leopold and MacGinitie, 1972.) Publication is authorized by the Director, United States Geological Survey. H.D.M.]

Pollen and Spores of the Kisinger Lakes Fossil Leaf Locality

by Estella B. Leopold

ABSTRACT

The Kisinger pollen and spore flora is a challenge to botanists; not only is it rich in species (161) and families (at least 41), but the total plant evidence indicates that many of the taxa are related to living groups of tropical distribution. Even though the leaf flora is large, the pollen and spore evidence adds some 16 additional genera and 14 families to the known flora. Additional potential for reconstruction of middle Eocene ecosystems may be derived from further study of this pollen flora. In this initial study, I was able to place only about one-sixth of the pollen-spore flora in living groups, but obviously additional information lies hidden in the unidentified fraction.

A rich mixture of angiosperm pollen including moist tropical and warm temperate types predominates in most samples, and suggests verification of the mesic and very warm conditions inferred from the megaflora.

The pollen and spore flora at and above the Kisinger leaf locality shows distinguishing characteristics that separate it from earlier Eocene floras regionally, and from younger (late Eocene) floras of western Wyoming. The flora, which by pollen-spore evidence is found over a wide area in central and northwestern Wyoming, is certainly of Bridger A–B age (early middle Eocene), but its stratigraphic span may include younger Bridgerian strata.

INTRODUCTION

The Kisinger pollen and spore flora is distinctive not only because it is rich in species and families, but because it includes a wide range of tropical and warm temperate groups and a few marker fossils characteristic of Bridgerian (middle Eocene) strata. The pollen flora is widespread over about 30,000 square miles of western and central Wyoming.

The plant megafossils of the Kisinger flora are described and documented by MacGinitie in this volume. This chapter is a study of pollen and spores of vascular plants found in sediments at and above the main Kisinger leaf locality in the Fish Lake and Kisinger Lakes quadrangles, northwestern Wind River Mountains, Fremont County, Wyoming.

PURPOSE AND SCOPE OF STUDY

The main purposes are to document the array of pollen and spores occurring at and above the Kisinger leaf beds and to compare taxonomically the pollen flora with the leaf flora. I have attempted to determine some of the affinities with living vascular plants that now can be established on the basis of a brief study of fossil pollen evidence. The different pollen and spore entities that could be found in 6 sediment samples were photographed by Bernadine Tschudy, and these are shown in plates 36 to 45 in this volume. We consider this an initial survey of the Kisinger pollen flora because there are still a number of types not yet photographed in this large flora. General comparisons of the pollen and spores from the leaf locality with available Eocene collections from Wyoming have been made.

TAXONOMIC IDENTIFICATIONS

In preparation for identification of fossil pollen and spores, I examined modern pollen or spores for all of the plant genera identified from the Kisinger leaf flora by MacGinitie (this volume). This effort suggested a number of identifications for the fossil material. Each fossil pollen or spore type was compared with the photographic file of modern reference material at the United States Geological Survey pollen laboratory in Denver, in a search for taxa not included in the leaf flora. Photographs of fossil material were then compared with modern pollen slides and the modern material was compared directly with the fossils.

The systematic list of orders, families, and genera based on pollen and spore evidence (p. 53) should be considered a beginning. The list indicates definite generic and family identifications where possible. Where the fossil pollen or spores structurally resemble modern groups but the resemblance does not permit a definite assignment, a suggestive identification is indicated by the use of "cf." (= *con forme* meaning "compare with"); in this chapter, "cf." is merely permissive evidence that the form might belong to the cited group. For a large number of palynomorph types, I have as yet no real suggestions as to modern affinity.

The coordinates given in the plate descriptions are from the stage of the Zeiss photomicroscope at the United States Geological Survey pollen laboratory in Denver, Colorado. For conversion of our coordinates to those of another mechanical stage: our coordinate readings for the center point of a 1×3 inch standard microscope slide are 108.2 (horizontal axis) $\times$ 12.4 (vertical axis). When the slide label is placed on the left on the microscope stage our vertical coordinates decrease toward the lower edge of the slide.

The locality numbers are permanent United States Geological Survey Paleobotanical locality numbers and these are used as slide numbers with the addition of the number in parentheses (see plate descriptions).

The plate figures are all shown at the same magnification except those on plate 38 which are reduced because of the large size of the specimens.

LOCALITIES AND AGE OF SAMPLES

Pollen and spores illustrated in the plates are from six sediment samples collected by W. L. Rohrer at and above the main leaf-bearing horizon of the Kisinger flora ("A" Horizon of Rohrer, this volume; see p. 16); these samples span about 700 feet of sediments of the upper volcaniclastic unit and the lower part of the transition zone (see Rohrer's chapter in this volume, figs. 3, 4, p. 10, 12). I consider these samples to be Aycross Formation as did Leopold and MacGinitie (1972) (see fig. 8). The sampled stratigraphic interval is bracketed by radiometric age determinations of 49 and 46 million years (Rohrer, this volume); the pollen material is therefore of middle Eocene age. The 6 pollen samples are listed below in stratigraphic order, youngest at the top:

USGS PALEOBOT. LOC. NO.

D3530

TRANSITION ZONE SEDIMENTS

Elev. 10,020 ft, C SE¼ Sec. 4 (unsurveyed), T 43 N, R 109 W, 6th PM, 180 ft above base of formation; from basal part of a 1 ft coal bed and the underclay. About 830 ft above the White Pass bentonite (49.3 m.y.).

D3531A   Elev. 9,840 ft, C NE ¼ NE ¼ Sec. 4 (unsurveyed) T 43 N, R 109 W, 6th PM from lower part of a coal bed (probably a continuation of above), and from same (?) stratigraphic position as above.

D3531B   Data as above except: from a thin coal streak 6 ft below sample D3531A.

D3531C   Data as above except: sample from 1 inch below sample D3531B.

UPPER VOLCANICLASTIC UNIT

D3532A and D3532B   "A" flora horizon, elev. 9,100 ft. N½S½S½ Sec. 12 (unsurveyed), T 43 N, R 109 W; 270–350 ft above the bentonite of White Pass. These samples are about 500 ft stratigraphically below localities D 3531 and D3530.

The pollen flora of these samples from the main Kisinger leaf locality was compared with pollen from some 40 other Eocene pollen collections from Wyoming; 6 of these collections contain assemblages that resemble the Kisinger pollen flora. They include the following two localities from the Kisinger Lakes quadrangle where Rohrer found outcrops of the upper volcaniclastic unit containing the same Kisinger leaf flora:

USGS PALEOBOT. LOC. NO.

D4395A—From the "A" flora horizon, basal shale, black, coaly sediments 1.8 ft in thickness; N½NE¼NE¼ Sec. 5, T 43 N, R 109 W, Kisinger Lakes quad., Fremont Co., Wyo.

D4396—From "A" flora horizon, SE¼NW¼ Sec. 9, T 43 N, R 109 W, about one mile SSE from D4395.

The other 4 localities are from various formations in the western half of Wyoming, as follows:

USGS PALEOBOT. LOC. NO.

D3321—Bridger Formation, NW¼NE¼ Sec. 25, T 30 N, R 106 W, Sublette Co., Wyo. "Big Sandy Bend" locality in the northern part of the Bridger Basin, near Big Sandy, Wyo. in drainage of the Green River; collected by J. D. Love. Pollen flora closely similar to that at Kisinger leaf locality.

     Sample A—Leaf bed directly overlying a vertebrate fossil bed of early Bridger B (early middle Eocene) faunal age (West, 1969: 87–88).

     Sample B—Same locality as A but seed bed at water level, downstream from stock bridge and below the vertebrate fossil bed.

     Sample C—Same locality as A, ostracode-bearing shale in lower Bridger below the vertebrate fossil bed, downstream from bridge, near water level.

D3895—Aycross Formation, type area Wind River Basin, SW¼ Sec. 5, T 42 N, R 105 W, Fremont Co., Wyo.; collected by J. D. Love. An early middle Eocene age, based on vertebrate evidence, is indicated (Love, 1939). Pollen flora closely similar to that at Kisinger leaf locality.

D3795—Wagon Bed Formation, SE½ Sec. 25, T 33 N, R 95 W, Fremont Co., Wyo., about 10 miles southeast of Riverton. Collected by R. A. Scott. Sample is from Van Houten's (1964) Unit 3 which elsewhere is of Bridgerian age (Lohman and Andrews, 1968). Pollen flora extremely similar to Kisinger flora. The pollen assemblage from this sample is listed by Leopold and MacGinitie (1972), who termed it of early Bridgerian age.

D4328—Green River Formation, basal part of Laney Member from a tuff layer; southern edge of Green River Basin, "Blacks Fork" 10 miles SW of Green River, Wyo., on east side of cut, north of bridge. SE¼, Sec. 8, T 16 N, R 108 W, Sweetwater Co., Wyo. Sample collected by H. D. MacGinitie. Pollen flora generally similar to Kisinger flora. From its stratigraphic position in the Green River Formation, the age is early Bridgerian (West, 1969: 91).

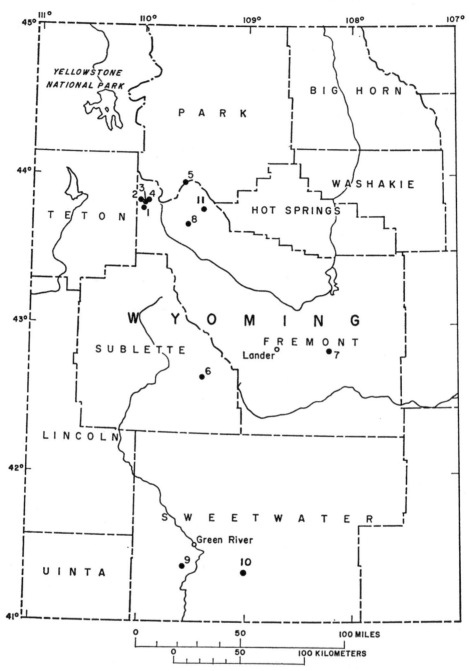

Fig. 7. Map showing USGS paleobotanical localities mentioned in text. 1, D3530, Transition zone; 2, D3531, Transition zone; 3, D3532, Kisinger "A" floral horizon, main leaf locality; 4, D4395A, D4396, Kisinger "A" floral horizon; 5, D4617, Wiggins Formation; 6, D3321, Bridger Formation, "Big Sandy Bend"; 7, D3795, Wagon Bed Formation; 8, D3895, Aycross Formation, type area; 9, D4328, basal part of Laney member, Green River Formation; Blacks Fork locality; 10, D4312, basal part of Laney Member, Green River Formation; 11, D3894, Tepee Trail Formation, type section.

D4312—Green River Formation, basal part of Laney Member; Washakie Basin NE¼ Sec. 21, T 16 N, R 100 W, Sweetwater Co., Wyo. (Pollen flora generally similar to the Kisinger pollen flora.)

The geographic location of the pollen samples in Wyoming is shown on the index map (fig. 7). No samples from the Tipperary leaf locality were available for pollen analysis. Stratigraphic usages by various authors are shown in figure 8. Probable stratigraphic relations of samples discussed are given on figure 9.

PREPARATION OF FOSSIL SAMPLES

Surfaces of the sediment chunks were cleaned with a wire brush and soapy water, then broken up in a mortar. An initial treatment with weak HCl to remove carbonates was followed by a 2-hour exposure to HF under agitation. Organic material was floated off on heavy liquid ($ZnCl_2$ of density 2.1) after centrifuging 20 minutes at 1500 rpm. Schultze solution made with $KClO_3$ and concentrated nitric acid was then added to the sample and allowed to react for 25 minutes. The remaining residues were treated with acetolysis and cleaned by panning (swirled in a watchglass with water). Using differential centrifugation, some of the smaller than pollen-sized fines were poured off. After staining in Bismark Brown, the pollen was mounted in AYAF (vinyl resin) and sealed. These preparation methods are included in Gray (1969).

SYSTEMATIC LIST OF ORDERS FAMILIES AND GENERA BASED ON FOSSIL POLLEN AND SPORES

Equisetaceae ?
   cf. *Equisetum*
Lycopodiaceae
   Lycopodium cf. *L. clavatum**
Selaginellaceae
   Selaginella cf. *S. densa**
   cf. *Selaginella conduplicata**
Polypodiaceae
   cf. *Pteris**
Schizaeaceae
   Lygodium cf. *L. kaulfussi* Heer*
Osmundaceae
   Osmunda
Ginkgoaceae or Cycadaceae
Podocarpaceae ?
   cf. *Dacrydium**
Pinaceae
   *Pinus**
   cf. *Keteeleria**
   cf. *Cedrus*
   *Picea**
   *Tsuga**

Taxodiaceae
   Glyptostrobus or *Cryptomeria*
Taxodiaceae-Cupressaceae-Taxaceae*
Sparganiaceae
   Sparganium (not figured)
Gramineae
   cf. Palmae*
   cf. Lemnaceae
   cf. Liliales or Amaryllidales*
Salicaceae
   Salix
   cf. *Populus*
Juglandaceae
   *Carya**
   Pterocarya
   cf. *Juglans*
   cf. *Engelhardtia* (not figured)
Betulaceae
   *Alnus**
Fagaceae
   cf. *Castanea**

* Also found in local tuffaceous Wind River Formation (lower Eocene) at USGS Paleobot. locs. as follows (see stratigraphic discussion of Wind River Fm. p. 5):
   D3533, One Mile Creek flora, NW¼SE¼NW¼ Sec. 29, T 42 N, R 109 W, Fremont Co., Wyo.
   D3534, Coyote Creek flora, N½SW¼NE¼NW¼ Sec. 19, T 42 N, R 109 W, Fremont Co., Wyo.
   D3535 and D1873 Du Noir (Wind River) flora, SW ¼, SW ¼ Sec. 3, and N½ NW ¼ Sec. 10, T 42 N, R 108 W, 6th P.M., Fremont Co., Wyo.

Ulmaceae
  Trema
  cf. *Zelkova**
Eucommiaceae
  Eucommia cf. *E. ulmoides**
Olacaceae
  Anacolosa type (not figured)
Chenopodiaceae-Amaranthaceae
cf. Platanaceae
  cf. *Platanus**
Rosaceae? cf. *Prunus*
Leguminosae
  cf. *Mucuna*
  cf. *Psoralea*
  Bauhinia cf. *B. congesta*
Meliaceae
  Cedrela cf. *C. mexicana*
Polygalaceae
  cf. *Securidaca** (not figured)
Aquifoliaceae
  Ilex
Aceraceae
  Acer cf. *A. palmatum**
  Dipteronia cf. *D. sinensis*

Sapindaceae? (or Elaeagnaceae?)
  cf. *Serjania* (or *Elaeagnus*)
Tiliaceae
  Tilia
  Luehea
  *Triumfetta**
  cf. *Schoutenia**
  cf. *Apeiba*
Malvaceae
Bombacaceae
  Bombax
  Bernoullia
Sterculiaceae? Section Helictereae*
Alangiaceae
  Alangium†
Haloragidaceae? cf. *Gunnera**
Myrtaceae or Sapindaceae*
  cf. *Eugenia*
Sapotaceae
Rubiaceae*

An Angiosperm pollen of unknown affinity
  Pistillipollenites mcgregorii Rouse*

DESCRIPTION OF POLLEN AND SPORE FLORA

The general characteristics of the Kisinger pollen flora are its great diversity, strong taxonomic representation of the Tiliaceae and Aceraceae, and presence of groups, like Bombacaceae, which have limited stratigraphic range in the Rocky Mountains. It differs from older and younger Eocene floras as discussed under "Stratigraphic Implications."

SIZE

The pollen and spore flora of the Kisinger leaf locality embraces some 161 species, including 23 ferns or fern allies (pls. 36, 37), 10 Gymnospermae (pls. 38, 39), 7 Monocotyledonae (pl. 39), and 121 Dicotyledonae (pls. 39–45). Of these only 25 species can now be assigned to modern genera, 16 to families and 5 to groups of families. A total of 27 families and 24 genera are here identified with certainty, which means that modern analogues have so far been found for only about a sixth of the Kisinger pollen and spore flora. One measure of the richness and complexity of the flora is the diversity of pollen contained in single samples. In 5 out of 6 individual sediment collections, a hundred count yielded 35 or more species of pollen and spores. Early Eocene (Wasatchian) and middle Eocene (Bridgerian) pollen assemblages are comparatively rich in species, but these contrast with the the simpler (smaller) floras of late Eocene (Uintan) age.

---

* Also found in local tuffaceous Wind River Formation (lower Eocene) at USGS Paleobot. locs. as follows (see stratigraphic discussion of Wind River Fm. p. 14):
  D3533, One Mile Creek flora, NW¼SE¼NW¼ Sec. 29, T 42 N, R 109 W, Fremont Co., Wyo.
  D3534, Coyote Creek flora, N½SW¼NE¼NW¼ Sec. 19, T 42 N, R 109 W, Fremont Co., Wyo.
  D3535 and D1873 Du Noir (Wind River) flora, SW ¼, SW ¼ Sec. 3, and N½ NW ¼ Sec. 10, T 42 N, R 108 W, 6th P.M., Fremont Co., Wyo.
  † Found at USGS Paleobot. loc. D3321A

Composition

The flora is overwhelmingly dicotyledonous, and shows a strong representation of Tiliaceae, Aceraceae, Leguminoseae and Liliales types. In terms of numerical representation in specimen counts, fern spores, especially *Lygodium* spores, are common to frequent. Pinaceae pollen is typically rare, but in one sample Pinaceae and *Lygodium* compose most of the assemblage. Dicot species are many but no single species is frequent in the slide tallies.

The Tiliaceae pollen types are mainly of the long axial reticulate tricolporate form, as in *Triumfetta* (pl. 44, figs. 1, 2), *Luehea* (pl. 44, figs. 12*a, b*) and cf. *Apeiba* (pl. 43, figs. 22, 23). A second and common type, probably related to *Schoutenia,* are the triporate round spiny pollen shown in plate 42, figs. 6, 7, 8. The brevicolporate spiny types (pl. 45, figs. 7, 8) may belong to Tiliaceae too. According to my observations, cf. *Schoutenia* pollen is present in the tuffaceous Wind River floral horizon below the upper volcaniclastic unit but is lacking from post-middle Eocene floras in the region. *Triumfetta* pollen is present in floras of early Eocene age but not in floras younger than middle Eocene in Wyoming.

Pollen of Aceraceae follow three distinctive patterns in this flora: (1) striate tricolporate longaxial pollen of the *Acer palmatum* type (pl. 43, figs. 4, 5, and perhaps figs. 6, 8); (2) striate tricolpate longaxial pollen of the *Acer pseudoplatanus* type (pl. 41, figs. 2, 3); and (3) obloid tricolporate, weakly reticulate pollen of *Dipteronia* (pl. 43, fig. 2–3). In Wyoming we have found pollen of pattern No. 1 in the Wind River (early Eocene) flora but not in floras younger than middle Eocene. Pollen of pattern No. 2 is common in younger Tertiary collections in the region. *Dipteronia* (pattern No. 3) so far appears to be confined to the Kisinger samples.

At least two pollen forms of the Kisinger flora may have restricted stratigraphic ranges in the Rocky Mountain region. One includes pollen of the tropical family Bombacaceae which in this area is so far known only from middle Eocene sediments (pl. 45, figs. 4, 5, 10). A second is the large unnamed tricolpate form shown in plate 40, fig. 9; it has striking thickenings along deep furrows and microreticulate sculpture. Both Bombacaceae pollen and this form of uncertain identity are rare but consistent members of lower middle Eocene floras of the Kisinger type in Wyoming. For example, both have been found in all the Bridgerian floras shown on the map in figure 7 (Wiggins and Tepee Trail Formations excepted), and so far both groups are unknown in older or younger regional strata.

Characteristic in the Kisinger flora are spores of the fern species *Lygodium* cf. *L. Kaulfussi* (pl. 36, fig. 10). These different-looking *Lygodium* spores are large, ca 90 mu, with thickened walls at end of and along laesurae. Those occurring in the matrix sediment are identical with spores which I removed from a sporangium of the megafossil *Lygodium kaulfussi* Heer—a fossil that MacGinitie has determined has close affinity with the living species *Lygodium palmatum* of southern Japan. *L. kaulfussi*-type spores (and leaves) are common to frequent in both the upper and lower volcaniclastic units and transition zone sediments and in the Wind River floral horizon.

ECOLOGICAL AND FLORISTIC ASPECTS OF IDENTIFIED ELEMENTS

As in the leaf flora, tropical or subtropical groups are well represented, *Lygodium*, cf. *Glyptostrobus* or *Cryptomeria, Bauhinia, Luehea, Triumfetta, Bombax, Bernoullia* and Sapotaceae. Some of these are characteristic members of the Evergreen Broadleaved Sclerophyllous Forest of southeast Asia (*Bauhinia*), or of Rain Forest (*Bombax* and *Glyptostrobus*); though *Glyptostrobus* is not definitely identified from pollen, it is present in the leaf flora.

At least two tropical genera of this pollen flora are confined to the New World (*Bernoullia, Luehea*).

Moist climate, warm temperate forms are numerous: *Carya, Tilia, Trema, Alangium, Dipteronia, Acer, Eucommia.* (*Alangium* pollen was not found at the Kisinger leaf locality, but it occurs at the Big Sandy Bend site D3321A.) These all occur in the Mixed Mesophytic Forest of the Yangtze Valley in southern China, though *Carya, Acer* and *Tilia* are wide-ranging warm temperate forms. One obvious member of the Mixed Mesophytic Forest, *Platycarya*, is conspicuous by its absence. In southeast Asia *Pterocarya* is restricted to the Mixed Mesophytic Forest, but has some isolated species in the Caucasus Mountains of southern Europe; there it thrives in a dry winter climate as a streamside tree.

Cosmopolitan genera, especially of the lower vascular plants and Gymnospermae compose a large part of the systematic list: *Lycopodium, Selaginella, Osmunda, Pinus, Salix, Alnus*, and *Ilex.*

Only two forms are restricted to cool temperate regions in their modern distribution—*Picea* and *Tsuga.* The rare pollen occurrences of these members of "boreal forest" in this flora suggest that coniferous boreal forest was not abundant locally but could have grown on nearby highlands. Wang (1961) points out that *Picea* and *Tsuga* now grow just above tropical forests in the high mountains of Taiwan. Both of these genera are represented in the higher part of the section sampled here (locality D3530) where they are associated with a rich assemblage of dicot pollen; other Pinaceae are poorly represented.

Many pollen and spore forms of this flora are identified only on the family level, or are of groups that have a generalized pollen morphology so that identification on a meaningful level is impossible. This means one cannot say on the basis of palynology what their ecological requirements might be. It may be significant that some 8 (or 33 percent) out of the 24 identified plant genera are most closely related to taxa that now grow in tropical or subtropical regions. Seven (or 29 percent) out of 24 identified genera represent groups that are now of warm temperate distribution. Cool temperate forms (2 out of 24 genera or 8 percent) are few.

TALLIES OF GENERAL POLLEN AND SPORE TYPES IN KISINGER SAMPLES

Relative abundance of general pollen and spore types in the samples is indicated by counts of a hundred grains as shown in accompanying table.

A feature of this flora is that a rich assemblage of dicots occurs in each sample (except for sample D3532A), and spores of *Lygodium kaulfussi* type are always present and usually abundant. No individual angiosperm species represents more than 15 percent of the count.

Angiosperm pollen represents from 37 to 89 percent of the pollen-spore count in

USGS Paleobot. locality number:	Kisinger "A" horizon		Transition zone, 500' above D3532			
	D3532			D3531		D3530
	A	B	C	B	A	
Lower plant spores						
Monolete spores	1	39	12	5	14	7
Lygodium cf. *L. kaulfussi*	30	6	13	44	9	1
Other trilete spores	2	–	4	14	2	2
Gymnospermae						
Disaccate pollen	55	10	1	–	3	–
Taxodiaceae-Taxaceae-Cupressaceae	–	1	4	–	1	–
Angiospermae						
Monosulcate pollen	–	1	2	–	4	–
Liliales-Amaryllidales	–	–	8	–	7	12
Inaperaturate pollen	1	2	5	5	13	2
Various dicotyledous pollen	11	40	51	32	48	75
Total pollen and spores:	(100)	(99)	(100)	(100)	(101)	(99)
Total percent angiosperms	(12)	(43)	(66)	(37)	(71)	(89)

five samples, but one collection (D3532A, the lowest stratigraphically) shows only about 12 percent angiosperms. In D3532A, disaccate pollen of the Pinaceae (mostly *Pinus*) dominate the count and are associated with an abundance (30 percent) of *Lygodium* spores, indicating that lowland forms are not entirely lacking.

RELATION OF POLLEN EVIDENCE TO KISINGER LEAF FLORA

Pollen and spore evidence definitely corroborates 12 of the 34 families identified from leaves, and suggests possible representation by pollen for 21 of the 34 families. (See table below comparing leaf and pollen floras.) Of the 35 modern plant genera witnessed by megafossils, only 7 can be identified with certainty from microfossils, but potentially equivalent microfossils are noted for 13 others.

No pollen evidence whatever was found for 10 plant families recorded by Kisinger leaves, probably for various reasons. For example, pollen of the family Lauraceae is not preserved in sediments, and fossil records of its pollen are unknown. Insect-pollinated groups such as Aristolochiaceae produce little pollen per flower and may release little pollen to the environment. Pollen structure in families such as Taxaceae, Taxodiaceae or Cupressaceae is so generalized that assignment to modern genera is difficult. Some groups present in the Kisinger leaf flora such as Hamamelidaceae, are wind-pollinated and are high producers of pollen; but megafossils of this family are, according to MacGinitie, very rare in the flora. Absence of Hamamelidaceae pollen may therefore be the result of the rarity of this group in the standing vegetation at the Kisinger site.

Pollen studies have added to the flora some 7 genera within the families identified by MacGinitie, and 14 additional families including 9 genera not found by him. The new additions include: *Pterocarya, Carya, Trema, Bauhinia, Acer, Tilia, Triumfetta, Selaginella, Osmunda, Pinus, Tsuga, Picea, Alnus, Eucommia, Bombax* and *Bernoullia*. The genera and families that evidence from pollen and spores have added to the flora are chiefly wind-pollinated groups, pointing out a heavy bias on the part of palynological records in favor of anemophilous plants.

CORROBORATION OF KISINGER LEAF FLORA BY POLLEN-SPORE EVIDENCE

FROM THE MEGAFOSSIL FLORA:	CORROBORATION BY THE MICROFOSSIL FLORA AND COMMENTS:
SCHIZAEACEAE	SCHIZAECEAE
Lygodium	*Lygodium*
PTERIDACEAE	
Acrostichum	
ASPLENIACEAE	
Asplenium	
THELYPTERIDACEAE	monolete spores
Thelypteris 2 spp.	(generalized structure)
SALVINIACEAE	
Salvinia	
EQUISETACEAE	EQUISETACEAE?
Equisetum	cf. *Equisetum*
TAXODIACEAE	TAXODIACEAE
Glyptostrobus	*Glyptostrobus* or *Cryptomeria*
CUPRESSACEAE	TAXODIACEAE-CUPRESSACEAE-TAXACEAE
Chamaecyparis	(generalized structure)
SPARGANIACEAE	SPARGANIACEAE
Sparganium	*Sparganium* found, not illustrated
PALMAE	PALMAE?
LEMNACEAE	LEMNACEAE?
Spirodela	cf. *Lemna*
SALICACEAE	SALICACEAE
Populus 2 spp.	cf. *Populus*
Salix	*Salix*
JUGLANDACEAE	JUGLANDACEAE
Juglans	cf. *Juglans*
	Pterocarya
	Carya
FAGACEAE	FAGACEAE
Castaneoides	cf. *Castanea*
ULMACEAE	ULMACEAE
Zelkova	cf. *Zelkova*
	Trema
ARISTOLOCHIACEAE	
Aristolochia	
LAURACEAE	(pollen of this family does not preserve
Laurophyllum	well)
HAMAMELIDACEAE	
Liquidambar	
PLATANACEAE	PLATANACEAE?
Platanus 2 spp.	cf. *Platanus*
ROSACEAE	ROSACEAE?
Prunus	cf. *Prunus*
LEGUMINOSAE	LEGUMINOSEAE
Caesalpinites	
Canavalia	
and 4 undet. taxa	
	Bauhinia
	cf. *Mucuna*
	cf. *Psoralea*
MELIACEAE	MELIACEAE
Cedrela	*Cedrela*

EUPHORBIACEAE
Acalypha
Aleurites
AQUIFOLIACEAE
Ilex
ACERACEAE
Dipteronia

SAPINDACEAE
Serjania
TILIACEAE
Apeiba
Luehea

STERCULIACEAE
Sterculia
DILLENIACEAE
Saurauia
MYRTACEAE
Eugenia
ARALIACEAE
Dendropanax
Schefflera
CORNACEAE
Cornus sp.
EBENACEAE
Diospyros
SYMPLOCACEAE
Symplocos

EUPHORBIACEAE?
cf. *Acalypha*

AQUIFOLIACEAE
Ilex
ACERACEAE
Dipteronia
Acer
SAPINDACEAE?
cf. *Serjania*
TILIACEAE
cf. *Apeiba*
Luehea
Tilia
Triumfetta
cf. *Schoutenia*
STERCULIACEAE?
Section Helictereae (not *Sterculia*)

MYRTACEAE?
cf. *Eugenia*

ADDITIONAL FAMILIES FROM POLLEN EVIDENCE
NOT APPEARING IN LEAF FLORA

SELAGINELLACEAE: *Selaginella*
POLYPODIACEAE
OSMUNDACEAE: *Osmunda*
GINKOACEAE or CYCADACEAE
PINACEAE: *Pinus, Picea, Tsuga* and 2 undet.
GRAMINEAE
BETULACEAE: *Alnus*

EUCOMMIACEAE: *Eucommia*
CHENOPODIACEAE-AMARANTHACEAE
MALVACEAE
BOMBACACEAE: *Bombax, Bernoullia*
SAPOTACEAE
OLACACEAE: *Anacolosa* type
POLYGALACEAE: *Securidaca* type

STRATIGRAPHIC IMPLICATIONS

In a broad zonation of Eocene leaf and pollen floras of Colorado and Wyoming, we (Leopold and MacGinitie, 1972) have described five major floristic phases based on known leaf and pollen floras of the region. Our sequence includes floras of early Eocene (Wasatchian provincial age) through late but not latest Eocene (Uinta B) age. In this chronology, the Kisinger flora and its correlatives represent our "Phase 3" of early middle Eocene (early Bridger) age. (see fig. 8)

The Kisinger pollen flora spans some 500 feet of section at and above the main leaf locality and appears at the same stratigraphic horizon ("A" flora horizon) several miles away, showing local consistency and stratigraphic continuity of the assemblage. The occurrence of essentially the same assemblage at three widely

Epoch	Provincial Age Terminology, North America Wood et al (1941)	Leopold and MacGinitie (1972) Leopold (this chapter) Main floristic phases, Wyo.	Rock units — Washakie Basin, Wyo.	Rock units — Wind River Basin, Wyo.	Rohrer (this volume) Kisinger Lakes quad, Wind River Basin, Wyo.	Keefer (1957) Du Noir area, Wind River Basin, Wyo.
Upper	Uinta	Phase 5	? Washakie Fm.[1] ?			
EOCENE Middle	Bridgerian	Phase 4 / Phase 3 / Phase 2	? Green River Fm. and tongues of Wasatch Fm. ?	? Aycross Fm. ?	? Wiggins Fm. ? / Transition zone / Upper volcaniclastic unit[2] (Tepee Trail Fm)	Tepee Trail Fm.
Lower	Wasatchian	Phase 1	Wasatch Fm.	Wind River Fm.	Lower volcaniclastic unit / Variegated unit ? (Wind River Fm)	Bentonite at White Pass / Wind River Fm.[3]

Fig. 8 Stratigraphic chart showing general relations of rock units and floristic phases mentioned in this chapter.

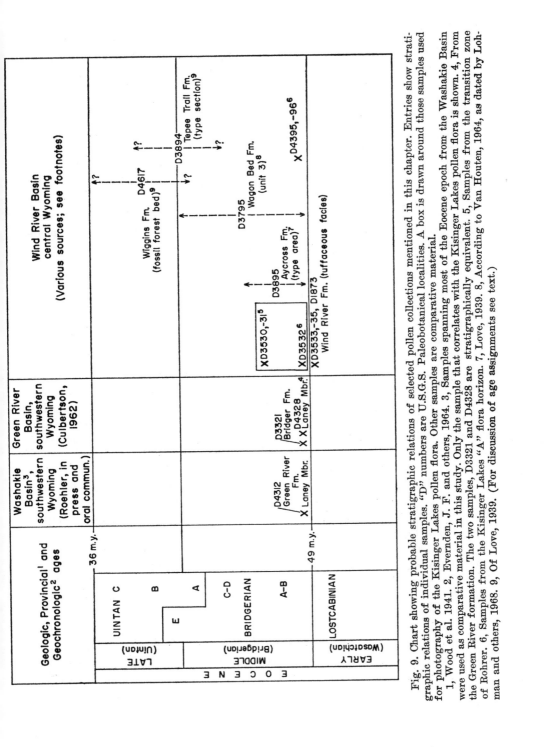

Fig. 9. Chart showing probable stratigraphic relations of selected pollen collections mentioned in this chapter. Entries show stratigraphic relations of individual samples. "D" numbers are U.S.G.S. Paleobotanical localities. A box is drawn around those samples used for photography of the Kisinger Lakes pollen flora. Other samples are comparative material. 1, Wood et al. 1941. 2, Evernden, J. F. and others, 1964. 3, Samples spanning most of the Eocene epoch from the Washakie Basin were used as comparative material in this study. Only the sample that correlates with the Kisinger Lakes pollen flora is shown. 4, From the Green River formation. The two samples, D3321 and D4328 are stratigraphically equivalent. 5, Samples from the transition zone of Rohrer. 6, Samples from the Kisinger Lakes "A" flora horizon. 7, Love, 1939. 8, According to Van Houten, 1964, as dated by Lohman and others, 1968. 9, Of Love, 1939. (For discussion of age assignments see text.)

different sites in separate stratigraphic units, each of established Bridgerian or early Bridgerian age, indicates that the Kisinger flora has regional stratigraphic significance.

In the discussion that follows, I have compared the Kisinger pollen and spore flora with other pollen assemblages that in various degrees resemble it. First are those localities at which the floras are essentially identical with the one reported here. Some localities are from the same stratigraphic unit, whereas others are from different formations. They are chiefly of known middle Eocene (Bridgerian provincial) age, established by fossil vertebrates and diatoms. I consider these to be stratigraphically equivalent to our Phase 3 floras. Then are cited Eocene floras that are generally similar to the Kisinger pollen-spore flora. Finally the Kisinger pollen flora is contrasted with assemblages that are quite different and are slightly older or younger than the Kisinger leaf beds. The general stratigraphic relations of the comparative pollen localities are shown in the chart in figure 9.

DISTRIBUTION OF POLLEN FLORAS SIMILAR TO KISINGER FLORA

Assemblages essentially identical to the pollen-spore flora from the main Kisinger leaf locality have been found at the same stratigraphic horizon ("A" horizon) in the same general area (Kisinger Lakes-Esmond Park quadrangles); the localities (D4395, D4396*) lie only a few miles to the northeast in the Wind River Basin.

Pollen floras practically identical to the Kisinger flora are also found in three different formations at far-flung localities in western Wyoming. One locality of middle Eocene age is in the middle part of the Aycross Formation at its type area (D3895), which is in the Wind River Basin but lies southeast of the Kisinger leaf locality. Vertebrates of two ages are reported from this section—a lower fauna considered of early Bridgerian provincial age (Bridger A), and an upper one similar to Bridger C faunas (Love, 1939). The pollen horizon lies stratigraphically between the vertebrate levels.

A second site at which a Kisinger-type pollen flora occurs is in the lower part (Unit 3) of the Wagon Bed Formation (D3795) along the southeastern rim of the Wind River Basin. The sample horizon (Unit 3) is of Bridgerian age (Van Houten, 1964; Lohman and Andrews, 1968).

A third locality recording a close analogue of the Kisinger pollen flora is the "Big Sandy Bend" site in the Bridger Formation at the northern edge of the Green River Basin (D3321). Inasmuch as the samples are from just above and just below a Bridger B vertebrate fauna (West, 1969), the flora at this site is established to be of early middle Eocene age.

Pollen floras somewhat similar to the Kisinger flora are found in southern Wyoming. From a survey of pollen samples from sections representing most of the Eocene in the Washakie Basin and from the southern part of the Green River Basin, I conclude that we have as yet not found a close analogue to the Kisinger pollen-spore flora in these southerly localities. Within this sequence pollen assemblages generally similar to the Kisinger flora are from the basal Laney Shale Member of the Green River Formation (i.e., D4328, Blacks Fork locality; and

---

* USGS Paleobot. loc. no.
  D4395 N½ NE¼ NE¼ Sec. 5, T43N, R109W., Wyo.
  D4396 SE¼ NW¼ Sec. 9, T43N, R109W., Wyo.

D4312, both of early Bridgerian age). The northern end of the Green River Basin (see loc. D3321) may mark the southern limit of the Kisinger type of flora.

An even more southerly stratigraphic equivalent to the Kisinger floral beds may be the Mahogany Ledge unit, Parachute Creek Member, Green River Formation from the Piceance Basin of northern Colorado (Henry Roehler, written commun., May 8, 1972). Pollen (in thin section slides) from this unit in Garfield County, Colorado, described by Wodehouse (1933) were studied and found to bear only a small resemblance to the Kisinger assemblage reported here.

RELATION TO OLDER EOCENE POLLEN FLORAS OF WYOMING

The Kisinger pollen-spore flora differs generally from older (early Eocene) floras of the region since it contains a somewhat larger proportion of tropical groups and shows poor representation of taxa characteristic of lower Eocene floras. It also differs from slightly younger middle Eocene floras from western Wyoming, as will be discussed here. In addition, it is very different from pollen assemblages of late Eocene (Uintan) age from southern Wyoming (Washakie Basin) where the floras are species-poor, and show a dominance of temperate families not well represented in the Kisinger flora.

For comparison with a lower Eocene flora I utilized sediments from the lower volcaniclastic unit (= upper part or tuffaceous facies of the local Wind River Formation as used by Keefer, 1957) that lies below the Kisinger A floral horizon. It should be noted that in this chapter I am using Keefer's term "Wind River Formation" for the lower Eocene sediments at and below the bentonite of White Pass. For some distance in the White Pass area some 12 miles east of the Kisinger flora site (D3532), this bentonite marks the local top of the Wind River Formation. The bentonite according to Rohrer is widespread and is a useful marker bed in the Kisinger Lakes quadrangle. As used here, the Wind River Formation includes the lowermost part of the upper volcaniclastic unit proposed by Rohrer (this volume) and underlying strata (see fig. 8). The beds are of late early Eocene (Lostcabinian) age.

Samples of Wind River Formation from localities D3533, D3534, D3535 and D1873 (locations given on p. 53) from a few hundred feet below the bentonite of White Pass were studied for their contained pollen flora. Only half of the Kisinger pollen and spore forms figured here are found in these local Wind River strata; the forms appearing in both floras are noted with asterisks in the list on page 53.

Characteristically abundant in the Wind River and in other lower Eocene strata of Wyoming are *Platycarya* (up to 28 percent of the pollen count) and *Pistillipollenites,* but these are rare (or lacking in the case of *Platycarya*) in the Kisinger floral beds. Some Wind River taxa not occurring in the Kisinger flora include *Ephedra, Abies, Bursera,* cf. *Pistacia,* large tetrads of Ericales, as well as a number of unnamed forms. In the local Wind River flora members of the Tiliaceae and Aceraceae are poorly represented (see list, p. 53) and Bombacaceae pollen is apparently absent. Hence, in the transition between the lower Eocene Wind River flora and the Kisinger flora of middle Eocene age, there is a decreased abundance of *Lygodium, Pistillipollenites,* a loss of *Ephedra, Abies, Bursera,* cf.

Pistacia, Ericales and other forms, and an appearance of Bombacaceae, and more members of the Tiliaceae and Aceraceae.

Pollen and spore tallies of the local Wind River collections show occasional samples with a high count of Pinaceae pollen and/or fern spores, but most are dominated (as in the Kisinger flora) by a very diverse assemblage of angiosperm pollen, no single species of which is very abundant.

I have studied pollen samples from the main body of the Wind River Formation below the tuffaceous facies elsewhere in the Wind River Basin, and also from other units of early Eocene age in Wyoming. The sequence of early Eocene pollen in this region is complex and has not been described to date; but enough is known about it to suggest that large changes occur within it. In none of the sections examined did the pollen-spore assemblages resemble the Kisinger flora and the resemblance was even less than in the suite from the local tuffaceous facies of the Wind River.

RELATION TO YOUNGER EOCENE POLLEN FLORAS OF WYOMING

Comparison of the Kisinger pollen-spore flora with slightly younger assemblages necessitates utilizing scattered localities from northern Wyoming, and with Uintan (late Eocene) sequences of pollen obtained in southernmost Wyoming from the Washakie and Green River Basins.

The northern localities are two, and include samples from the Wiggins and the type section of the Tepee Trail Formation. Neither flora resembles the Kisinger flora.

The type section of the Tepee Trail Formation is about 26 miles east of the Kisinger floral site (D3894, Section 4, T 104 W, R 43 N, East Fork Basin quad, Fremont Co., Wyo.). New vertebrate evidence provided by Malcolm McKenna (oral commun. Sept. 1, 1972) indicates that the type section contains more vertebrate forms diagnostic of Uintan than of Bridgerian provincial age. However, the K-Ar dates from the basal part of the type Wiggins Formation which overlies the Tepee Trail along the East Fork of the Wind River, is more than 47 m.y. in age (Obradovich *in* Smedes and Prostka, 1972., USGS Prof. Paper 729C, p. C 32). This indicates the considerable discrepancy between the middle and upper Eocene terminology as used in the vertebrate and K-Ar calendars.

Leaf collections from the Tepee Trail type section and vertebrate locality yielded a small flora including *Sapindus* similar to the living *S. drummondi* of Texas, Arizona, and southeastern states. It also includes abundant leaves of *Platanus* similar to *P. wyomingensis* from the Colorado Green River flora but a more advanced form with smaller leaves. One specimen of *Populus* is similar to or identical with the *P. wilmattae* of the Green River flora. The fossil leaves according to MacGinitie (written commun., Sept. 7, 1972) are of small size and suggest a riparian local environment in a climate warmer and drier than that of the Green River flora. Pollen of a single sample (D3894) shows a dominance of *Alnus,* with cf. *Platanus, Eucommia,* Rosaceae undet., *Acer, Abies?* and *Pinus.* The types of vertebrates verify that the site was a local pond or swale, but the pollen rain indicates influx of forms from drier uplands. In conclusion, though the flora is small, it suggests a hot, dry climate and is probably younger than the Green River flora.

The fossil forest locality of the Wiggins Formation lies some 25 miles northeast of the Kisinger leaf locality. It may be of middle and/or late Eocene age in the K-Ar calendar but probably is late Eocene in the vertebrate calendar. It is stratigraphically higher and younger than the Kisinger horizon. The forest bed (D4617A; Section 32, T 46 N, R 106 W, unsurveyed; projected grid, Emerald Lake quad., Fremont Co., Wyoming) contains a rich suite of coniferous wood types, and only one angiosperm wood (a ring porous specimen) has been found (R. A. Scott, oral commun., Nov. 12, 1973). Fossil pollen from the site is mainly *Pinus*, and aside from *Alnus* and *Pterocarya* contains little in the way of angiosperms. The presence of *Picea* and the abundance of *Pinus* suggests a montane coniferous forest under a cool temperate climate.

In summary, the floras of the type section of the Tepee Trail and the Wiggins Formation as far as they are known differ from each other and are not at all like the Kisinger flora described in this volume.

The final comparison of the Kisinger pollen flora is with late Eocene (Uintan Provincial age) pollen floras of southern Wyoming. The general character of Uintan A, B and C assemblages from the Washakie Basin is utterly different from that of the Kisinger flora; the flora is limited, being composed of cosmopolitan or definitely temperate groups. The prominent families in the Uintan are Pinaceae (mainly *Pinus*), *Ephedra* (several species), Gramineae, Rosaceae, Chenopodiaceae-Amaranthaceae, and a few hardwoods of temperate character: *Pterocarya, Eucommia, Ulmus-Zelkova,* and finally an abundance of aquatics: *Sparganium, Lemna* type, and Cyperaceae. I conclude that the late Eocene floras locally are improverished, temperate and of a transitional character between the rich, wet, mainly tropical aspect of the Kisinger flora and the warm temperate early Oligocene flora of the Florissant lake beds of central Colorado.

CONCLUSION AND SUMMARY

The Kisinger pollen and spore flora contains a diverse array of Eocene angiosperms. Rich in species (161) and families (at least 41), the assemblage has many taxa that are related to groups of tropical or subtropical distribution. Even though the leaf flora is large, the pollen and spore evidence adds some 16 additional genera and 11 families to the known flora. Further taxonomic study of this pollen flora could hold potential for reconstruction of local ecosystems, but from this initial study, I was able to determine modern affinities for only about one-sixth of the pollen and spore flora.

The rich mixture of dicotyledonous pollen, including moist tropical and warm temperate plant types, predominates in most samples. This evidence suggests verification of the mesic and very warm conditions inferred from the megaflora. Since pollen rain in areas of some topographic relief tend to reflect pollen of vegetation zones from various elevations, temperate forms may be expected to be more obvious in the pollen and spore flora than in the megaflora.

An abundance of pine pollen in the lower part of the Kisinger "A" floral horizon suggests that coniferous forest types were in the area, but the concomitant abundance of *Lygodium* spores and local absence of cool temperate conifers (*Picea, Tsuga*) suggest that the environment of deposition was of lowland rather than highland character, and that it was warm, not cool. The presence of cool temperate types (*Picea, Tsuga*) as occasional grains in the transition zone sediments (D3530) suggests pollen drift from vegetation at higher elevations.

The pollen and spore flora at and above the main Kisinger leaf locality display singular characteristics that distinguish it from lower Eocene floras regionally, as well as from local collections that are younger in age (Wiggins and type section of Tepee Trail Formation), and from younger (late Bridgerian and Uintan) floras of the Green River Basins.

The Kisinger pollen flora occurs in sediments of established early middle Eocene (Bridger B) age at one site, and at three other sites known to be middle Eocene (Bridgerian) in age. Though the flora may actually range through the middle Eocene in northwestern Wyoming, pollen sequences from the Green River Basins suggest that its stratigraphic span may be limited to early middle Eocene sediments.

The geographic distribution of the flora, as measured by pollen assemblages that are closely similar to the pollen at the main leaf locality, is now known to include northwestern Wyoming. Pollen floras of the same age from the southern Rockies do not show a close similarity to the Kisinger flora, which suggests the existence of a north-south vegetation gradient in middle Eocene time.

ACKNOWLEDGMENTS

I am indebted to Bernadine Tschudy for photography of the fossil pollen and spore material and preparation of the plates. I thank David Love, Robert H. Tschudy, and Willis Rohrer for suggestions on the manuscript. I owe gratitude to Harry D. MacGinitie for guidance and salute his enthusiasm and care in the study of Eocene floras.

[This concludes Dr. Leopold's contribution to the paper. H.D.M.]

SYSTEMATIC DESCRIPTIONS

One of the chief difficulties encountered in assembling a paper of this kind is the preparation of adequate illustrations of the various types. Often it seems impossible to photograph satisfactorily the fine details of venation. I have tried to improve on this by making enlargements in many cases. This does give a better idea of the characters although it is not a perfectly satisfactory solution. Sometimes details of venation may be brought out by careful retouching, although this carries obvious dangers. If the retouching is done by the author, and not relegated to an artist who has little or no knowledge of paleobotany, errors can be rather satisfactorily avoided.

I have arranged the descriptions in alphabetical order according to genera. The "Engler and Prantl" system appears to depart from reality in many ways and there seems little point in following it. The alphabetical order is straightforward and easier to use.

The type numbers are those of the University of California, Museum of Paleobotany at Berkeley. I have used type names according to the definitions in the Florissant paper (MacGinitie, 1953:79–80). The term type is used instead of holotype.

DESCRIPTIONS OF MEGAFOSSILS
Family EUPHORBIACEAE
Genus *Acalypha* Linné

Acalypha cirquensis, new species
(Pl. 26, figs. 1, 2; pl. 27, fig. 2)

Description.—Leaves ovate-lanceolate; length 6–15 cm.; width 2–5 cm., length/width ratio averaging 3.2; apex acute; base cuneate; margin irregularly serrate, the upper margin of the teeth perpendicular to the midrib; midrib slender; petiole stout, approximately ¼ the length of the lamina; 5 to 8 pairs of opposite to subopposite secondaries, originating at 40° to 45°, the basal pair acrodrome, arising at the top of the petiole and ascending for approximately ⅓ the length of the lamina, approaching within 4 to 9 mm. of the margin; secondaries strong, nonflexuous, pursuing a slightly curved course to just within the margin, bifurcating and sending one or more tertiary branches to the marginal teeth, the extremity of the secondary forming complex tertiary loops marginally; tertiary venation quadrangular-reticulate for the most part but with scattered, percurrent cross-ties; areolation a fine, quadrangular mesh of two orders, quartary meshes 0.12 to 0.15 mm. in size, quintary meshes 0.03 to 0.05 mm. in size; areolation closed; texture firm; frequency, common.

Discussion.—This species is similar to *A. serrulata* Potbury (1935:72, pl. 7, fig. 7) from Plumas County, California, which was later synonymized with *A. aequalis* (Lesquereux) MacGinitie (1941:137, pl. 32, fig. 7) from the Chalk Bluffs flora. The species from the Kisinger Lakes locality has a more prominently serrate margin and stronger acrodrome secondaries. Among living species the fossil leaves are most like the foliage of *A. erosa* Rusby of southern Bolivia and *A. salicioides* Rusby from Colombia. The genus *Acalypha* comprises about 400 species of shrubs or small trees widely distributed in the tropics around the world. Some of the species have pronounced adaptations to subhumid conditions.

Occurrence.—Kisinger Lakes, Tipperary. Type PA 5705, paratype PA 5707, topotypes PA 5708, PA 5709.

Family POLYPODIACEAE
Genus *Acrostichum* Linné
Acrostichum hesperium Newberry
(Pl. 19, fig. 3)

Acrostichum hesperium Newberry, U.S. Geol. Surv. Mon. 35, p. 6, pl. 61, figs. 2–5, 1898; Mac-
Ginitie, Univ. Calif. Publ. Geol. Sciences, vol. 83, p. 89, pl. 4, fig. 1, pl. 12, fig. 2, 1969.

Although the figured specimens given by Newberry are somewhat misleading—the lateral veins
are not prominent enough—it appears that his identification is valid. In the Green River paper
cited above I synonymized *Musophyllum complicatum* Lesquereux. It now appears that the cross-
nervilles in this latter form are nearly perpendicular to the laterals and not at a high angle as
in *Acrostichum*. Thus the synonymy was an error and the true relationship of *Musophyllum* is
still unsolved. Fossils of *Acrostichum* are not uncommon in the Green River beds and at a
Bridger locality in the northeastern Wind River basin.

The fossil leaves are much like those of the living *A. aureum* Linné which is common on
tropical coasts around the world. The plant is extremely vigorous with fronds as much as 12
feet long. It forms nearly impenetrable jungles within and around the borders of mangrove
swamps. It can be seen in abundance in the swamp and lagoonal districts between Mazatlan
and Manzanillo, Mexico. It grows as far north as central Florida, where it is relatively common
in swampy hammocks.

Occurrence.—Kisinger Lakes, Tipperary, Valley. Hypotype PA 5677, topotype PA 5678.

INCERTAE SEDIS
Genus **Alafructus,** new genus
Alafructus lineatulus (Cockerell) MacGinitie
(Pl. 15, fig. 3)

Banksites lineatulus Cockerell, U.S. Nat. Mus. Proc., vol. 66, art. 19, p. 8, pl. 2, fig. 3, 1925.
Lomatia lineata (Lesquereux) MacGinitie, Carnegie Inst. Wash. Publ. 599, p. 108, pl. 34, figs.
    4–6, 1953. (Figs. 4–6 only.)
Lomatia lineatulus (Cockerell) MacGinitie, Univ. Calif. Publ. Geol. Sciences, vol. 83, p. 99, pl.
    16, figs. 6, 7, 1969.

Description.—Winged fruits (samaras) 9 to 14 mm. in length, 4 to 5 mm. in width. The seed
ovate, 4 mm. in length, 2.5 mm. in width, the long axis at an angle of 40° with the wing. Wing
short-ovate, the apex bluntly rounded, marked by 4 to 5 longitudinal veins, which curve upward
and coalesce at the end of the samara. The seed marked by a pronounced, crestlike projection
distally and a remnant of a thin, wiry pedicel proximally. Fruits borne singly, not paired or
multiple.

In a recent issue of Audubon Magazine (Moore and Ratcliffe, 1971:17) there is a photograph
of a fossil specimen, found at Florissant, Colorado, showing a branch with attached leaves and
many of these winged seeds. This shows clearly that these objects cannot be assigned to the
Proteaceae. They resemble the fruits of *Securidaca* (Polygalaceae) but the venation of the
wing is entirely different; the strong longitudinal veins, upcurving distally, are absent. It may
be that these fruits are those of some extinct genus of the Polygalaceae, but for the present
their true relationship must remain an unsolved problem. Therefore I have chosen to establish a
new genus *Alafructus* (winged fruit). The type is chosen as figure 6 in the 1953 publication
listed above. The fruits named *lineata* and *lineatulus* differ only in slight variations in size.
Dilcher, in his review of the Green River paper (1971:740) objected to the union of isolated
fruits and leaves under the same species, and these "Lomatia" fruits show that his point is well
taken.

This type of winged fruit occurs in the Florissant, Green River, Crooked River and Grant,
Montana, floras. The comparative rarity of the fruits and the common absence of associated
foliage, similar to that illustrated by Moore and Ratcliffe, suggests that the plant inhabited high
ground, perhaps frequenting dry slopes.

Occurrence.—Kisinger Lakes. Hypotype PA 5657, topotype PA 5659.

Family EUPHORBIACEAE
Genus *Aleurites* Forster
Aleurites fremontensis (Berry), new combination
(Pl. 21, figs. 1, 2; pl. 22, figs. 1, 2)

Aralia notata Berry, U.S. Geol. Surv. Prof. Paper 165 (no pg. ref.), pl. 12, fig. 6, 1930*b*.
Aralia notata var. *denticulata* Berry, *ibid.*, p. 75, pl. 15, fig. 5.
Ficus ungeri Lesquereux, Berry, *ibid.*, p. 70, pl. 12, fig. 4 only.
Negundo fremontensis Berry, *ibid.*, p. 72, pl. 11, figs. 1–3.

Description.—Leaves ovate, of variable size; length 3.5 to 14 cm., width 3 to 12 cm., length/width ratio 1.1 to 1.5; leaves trilobate, the conical central lobe extending about half the length of the lamina; apices of the lobes acute; sinuses open and rounded; petiole strong, averaging about ⅔ the length of the lamina; margin entire in the lower third, commonly irregularly dentate above, the dentations sharp and glandular, leaves occasionally nearly entire; venation prominently acrodrome with two pairs of primaries or pseudo-primaries arising from the top of the petiole or suprabasilar, the veins branching from the midrib within the lamina, the inner pair arising at 45°, pursuing a straight course to the apices of the lateral lobes, a second pair of weaker primaries branch off just within the margin at angles of 60° and follow close to the margin for about half the length of the lamina, looping with secondaries from the inner primaries; 3–4 pairs of subopposite secondaries arising at angles of 50°–60°, curved upward and approaching parallelism with the margin or branching and entering the marginal teeth where present, 5–6 secondaries arising from the outer side of the inner primaries, gently curved upward and commonly craspedodrome, where the margin is entire the secondaries form simple loops the outer elements of which form a marginal vein; tertiaries percurrent, perpendicular to the midrib and curved downward to become perpendicular to the secondaries; thin, quartary branches anastomose between the tertiaries, usually branching once, the quartary meshes of irregular size and shape but predominantly quadrangular; quintary venation a fine, closed mesh; the surface of the leaf characterized by fine dots or druses (calcium oxalate ?), 14–20 dots over the area of each quartary mesh; texture firm; frequency, common.

The original *"Aralia" notata* was created for a leaf of *Platanus nobilis* (Brown, 1962:64), and it appears that many platanoid leaves with three lobes have since received the epithet. Much of the material noted in La Motte's synonymy (1952:71) represents the foliage of *Platanus*. The minute punctations scattered over the surface of the fossil impressions of *Aleurites* are sufficient to distinguish them from the impressions of *Platanus* leaves. The fine, even-textured, quartary mesh, which is so characteristic of *Platanus* foliage is not found in the fossils. It may be said, that, in general, the resemblance of these fossil leaves to those of *Platanus* and *Aralia* is of the most superficial kind; there are practically no significant correspondences. The fossil leaves from the Wyoming localities are the impressions of the foliage of *Aleurites*, similar to that of the living *A. moluccana* (Linné) Wildenow of southeast Asia.

Two fossil species of *Aleurites* have been described from western North America: *A. americana* Potbury (1935:72, pl. 8, figs. 1, 4), and *A. glandulosa* (Brown) MacGinitie (1969:114, pl. 25, fig. 1). The first species is from the auriferous gravels of Plumas County, California, of early Oligocene age, and the second is from the Green River Formation of northeastern Utah, of Middle Eocene age. The fossils from Kisinger Lakes and Tipperary differ from those at La Porte in several respects; the lateral lobes are prominent with acute apices, the margin tends to be denticulate, and the tertiary venation is percurrent rather than reticulate. They bear considerable resemblance to the Green River species, but this fossil from northeastern Utah does not possess the long, attenuate middle lobe, nor the tendency for the greatest width of the lamina to be in the lower third, common in the Kisinger Lakes and Tipperary impressions. The differences may be merely varietal, but, until more material from the Green River beds is found it seems best to keep the species separate.

The existing species of *Aleurites* are warm temperate to tropical in their distribution: southeastern Asia from central, southern China southward to New Caledonia. About 6 species have been recognized. *A. moluccana* which most resembles the fossil species, appears to be confined to tropical habitats. *A. fordii*, the tung oil tree, is cultivated in Florida and southern China and will endure several degrees of frost.

Occurrence.—Tipperary, Kisinger Lakes. Hypotypes PA 5683, PA 5685, PA 5688.

INCERTAE SEDIS
Ampelopsis tertiaria Lesquereux

Ampelopsis tertiaria Lesquereux, U.S. Geol. Surv. Terr. Rept., vol. 7, p. 242, pl. 43, fig. 1, 1883; Berry, U.S. Geol. Surv. Prof. Paper 165, p. 73, pl. 11, fig. 15 pl. 14, fig. 4, 1930; Knowlton, U.S. Geol. Surv. Prof. Paper 131, p. 170, 1923.

As far as can be determined only two specimens of this species have been found. Knowlton changed the name from *Ampelopsis* to *Parthenocissus*, which change was discarded by Berry. Since the original identification cannot be correct the change is pointless. Both the original types are missing and cannot be rechecked. The locality for Lesquereux's specimen has also been lost. He gave it as "Green River, above the fish beds." I have searched repeatedly for this locality without success; it is probably that "Green River" of Lesquereux was at a different locality from that of the present town of Green River, Wyoming. He may have been referring to the fish beds west of Kemerer rather than those southwest of Green River.

The specimen cited by Berry probably did not come from the Tipperary locality. Berry, when he wrote his paper had not visited the place; the fossil material was collected by Mr. N. H. Brown of Lander, Wyoming, and contained a few specimens from Crowheart Butte, and Lenore, Wyoming (Berry, 1930b:55). These last two localities are not "Bridger" in age, but are of Lost Cabin or earlier age, and the "Ampelopsis" may have come from one of them.

Whatever the taxonomic position of this fossil may be it is not *Parthenocissus* or *Ampelopsis;* there are too may differences of form and venation. It is certain that the original fossil was 4- to 5-foliate as shown in the drawings (Lesquereux, 1883). The evenly and sharply serrate margin and the close-set, percurrent tertiaries resemble the foliage of *Platycarya,* a genus that has left abundant pollen grains in the early Eocene of the area, but the fossil shows that it was a palmately compound and not a pinnately compound leaf.

Occurrence.—Uncertain.

Family TILIACEAE
Genus *Apeiba* Aublet
Apeiba improvisa, new species
(Pl. 33, figs. 1, 2, 3)

Ficus wyomingiana Lesquereux, Berry, U.S. Geol. Surv. Prof. Paper 165, p. 70, pl. 12, figs. 1–3, 1930; *Ficus mississippiensis* (Lesquereux) Berry, mutant *pseudopopulus* Lesquereux, Berry, *ibid.,* p. 71, pl. 8, fig. 7.

Description.—Leaves ovate, of variable size, length 5 to 15 cm., width 2.5 to 13 cm., 1/w ratio 1.9; often slightly inequilateral; margin entire, apex blunt to acute; petiole stout, 1.7 to 2.3 cm. in length, slightly expanded at the leaf junction; venation acrodrome, 5 (or rarely 3) primaries from the summit of the petiole, the inner pair at 25° to 30° with the midrib, the outer pair at 40° to 64° with the midrib; margin marked by a bounding vein; the inner primaries extending in a gentle curve to the upper third of the leaf, anastomosing with the first (basal) secondary in a series of 5 or more loops; the outer primaries looping and merging with the margin in the lower third or fourth of the lamina; four pairs of subopposite secondaries, the first pair arising at 40° in the lower third of the lamina, ascending parallel to the inner primary and looped with succeeding secondaries, the upper 3 pairs of secondaries essentially parallel in looping and direction; tertiary venation percurrent with prominent cross-ties bridging the primaries and secondaries and approximately perpendicular to the secondaries, the tertiary veins often showing Y-branching, 4 to 5 prominent veins branch from the inner primaries and loop just within the margin; quartary venation mainly of thin, irregular cross-ties perpendicular to the tertiaries; the quartary meshes enclose a fine, even-textured, quadrangular mesh, the openings about 0.2 mm. in dimension; no finer details preserved; texture coriaceous; frequency, rare.

Discussion.—These fossil leaves, as Berry remarked, possess no true characters of *Ficus* foliage. He suggested their assignment to the Lauraceae, and pointed to resemblances with the leaves of *Oreodaphne* (= *Ocotea*) and *Sassafras.* The frequent occurrence of two pairs of primary veins, the peculiar quartary venation, and the pronounced marginal vein, set these fossil leaves aside from those of any existing Lauraceous genus. Their true affinity is with the foliage of the Tiliaceae, and, in that family, the pronounced correspondences are with the leaves of various species of *Apeiba*, especially *A. aspera* Aublet of tropical America. Berry also synonymized this species with Lesquereux's *Ficus*. There is no justification for this procedure. The only characters in common are the shape and three primary veins; otherwise the venation appears to be entirely different. The extended petiole and the details of venation illustrated suggest that Lesquereux' fossils might be the impressions of *Dendropanax* (*Gilibertia*) leaves. If so the term *wyomingiana* might be applied to the *Dendropanax* fossils discussed in this paper. However, I have not been able to locate the original types, and the locations given are extremely vague. The locality given by Lesquereux as "Evanston, Wyoming; Green River group" strongly indicates that both the *Ficus wyomingiana* and the *Ficus pseudo-populus* are from the Paleocene horizon near Evanston. Brown (1962:62–63) accepted this as true for *F. pseudo-populus*, but called the *F. wyomingiana* Eocene in spite of the fact that Lesquereux (1878:205) says "Green River group with the former" referring to *F. pseudo-populus*. The carelessness of the early paleobotanists with regard to exact locations and horizons has rendered useless some otherwise significant fossil types. Thus, since it does not seem possible to accurately identify "*Ficus wyomingiana*" at this time or to find the locality of its occurrence, it seems best to disregard the earlier described fossils until the original locality is rediscovered. Therefore I have ignored the species name *wyomingiana* and assigned a new species name.

There are about 7 species of *Apeiba* found in tropical America from southern Mexico through Costa Rica and Panama to South America. Plants of the genus are moderately sized trees which yield useful wood. They are deciduous or semi-deciduous members of the tropical rain forests or sub-deciduous forests.

Occurrence.—Tipperary, Kisinger Lakes. Type PA 5725, paratype PA 5726, topotype 5727.

Family ARISTOLOCHIACEAE
Genus *Aristolochia* Tournefort
Aristolochia solitaria, new species
(Pl. 8, figs. 1, 4)

Description.—Leaf ovate; length 5 to 7 cm.; width 3 to 4 cm.; greatest width at the base; margin entire; petiole not preserved; apex acute; base incipiently cordate; midrib slender, a pair of acrodrome veins arise from the top of the petiole at an angle of 30° and pursue a straight course until looping along the margin in the upper half of the leaf; 3 pairs of subopposite secondaries originate at 60°, curve upward parallel to the acrodome primaries and loop along the margin; 6 to 7 sub-secondary veins originate at large angles along the marginal side of the acrodrome primaries and form prominent loops just within the margin by means of anastomosing bifurcations; tertiary venation largely percurrent although with many Y-branches, veins aligned nearly perpendicular to the midrib, many tertiaries along the midrib bridging between this and the secondaries; quartary venation a prominent, polygonal mesh, the enclosing veinlets perpendicular to the tertiaries, the meshes of variable size but approximately 1 mm. in dimension; quintary veinlets of thin texture, the mesh angular-quadrangular, the openings 0.3 mm. in dimension; further details of venation not preserved; texture firm; frequency, rare.

Discussion.—This fossil clearly displays the characteristic venation of *Aristolochia* leaves. It closely resembles the foliage of *A. anguicida* Jacquemot, a strong growing vine widely distributed in the lowland and coastal regions of the West Indies and Central America. There are also significant resemblances to the leaves of *A. tagala* Chamisso from tropical southeastern Asia. The fossil does not show the strong, bifurcating basal secondaries common to many species of the genus. Berry (1924:59, pl. 9, figs. 7, 8) has described a species of *Aristolochia* somewhat similar to *A. solitaria*, from the Yegua Formation of the Gulf Coast, but this species from Louisiana lacks the prominent acrodrome venation of the Wyoming species. *A. mortua* Cockerell from the Florissant, Colorado flora (MacGinitie, 1953:109, pl. 35, fig. 3) and the Green River flora of northwestern Colorado (MacGinitie, 1969:100, pl. 11, fig. 2) is similar to *A. solitaria*, but lacks the characteristic, quadrangular, fourth order venation of this fossil.

There are approximately 300 species of *Aristolochia* distributed throughout the tropical to temperate regions of the world, with the exception of Australia. Most of the tropical species are vigorous lianas of moist lowland and streamside habitats.

Occurrence.—Kisinger Lakes. Type PA 5623, paratype PA 5625.

Family POLYPODIACEAE

Genus *Asplenium* Linné
Asplenium eolignitum Berry
(Pl. 5, fig. 1)

Asplenium eolignitum Berry, U.S. Geol. Surv. Prof. Paper 91, p. 167, pl. 11, fig. 3, 1916; U.S. Geol. Surv. Prof. Paper 165, p. 62, pl. 8, figs. 2–4, 1930.
Gymnogramma haydenii auct. non. Lesquereux, 1878; Lesquereux, U.S. Geol. Surv. Terr., vol. 8, p. 122, pl. 19, fig. 2, 1883.
Asplenium serraforme Berry, U.S. Geol. Surv. Prof. Paper 165, p. 62, pl. 8, fig. 1, 1930.

These two species are clearly identical; the first was founded on a fragment of the second. This fern genus is found at several localities in the Wind River basin and also in the Green River flora of northwestern Colorado and northeastern Utah (MacGinitie, 1969:88). The occurrences cover the age range from late Early Eocene to late Middle Eocene. Berry likened the fossils to the foliage of the living *A. serra* Langsdorf of tropical America. There are also close resemblances in the cases of *A. persicifolium* J. Smith and *A. prionurus* J. Smith, both from moist, tropical habitats in the Philippines. *A. serraforme* is closely similar to *Aneimia eocenica* Berry (1916:164, pl. 9, fig. 7, pl. 10, fig. 2, pl. 11, figs. 1–2) from the La Grange Formation at Puryear, Tennessee (early Middle Eocene), but without the presence of sori no identity can be established. This fern is rather uncommon at the Wyoming localities.

Occurrence.—Tipperary, Kisinger Lakes. Hypotype PA 5611.

Family LEGUMINOSAE

Genus *Caesalpinites* Saporta
Caesalpinites pecorae (Brown) MacGinitie, new combination
(Pl. 13, fig. 2)

Caesalpinia pecorae Brown, Journ. Wash. Acad. Sci., vol. 46, no. 1: 106, figs. 7, 9, 1956.

The specimen figured was discovered too late to be included in the tables or general discussion. It resembles the Green River and Bear Paw Mountains species so closely that it seemed best to include it in that species (MacGinitie, 1969:108). Since this fossil form is probably not a *Caesalpinia*, but should be more appropriately assigned to *Entada* cf. the living *E. polystachya* (Linné) Britton, the generic epithet has been changed to the noncommittal *Caesalpinites*

Occurrence.—Kisinger Lakes horizon at Ramshorn Peak, hypotype PA 5648.

Family LEGUMINOSAE

Genus *Canavalia* A. P. de Candolle
Canavalia diuturna, new species
(Pl. 10, fig. 4; pl. 12, figs. 1, 3; pl. 34, fig. 3)

Description.—Leaflets ovate, with a small degree of asymmetry, greatest width ⅖ of the total length above the base; width 2.3 to 9 cm., 1/w ratio 1 to 1.5; length 4 to 11.5 cm., margin entire; apex acute to rounded mucronate; base rounded to incipiently cordate; petiolule short, relatively slender, enlarged at juncture with lamina, faintly ridged, average length 5 mm.; midrib slender; 6 pairs of irregularly disposed secondaries, the basal pair opposite, the succeeding pairs usually alternate, originating at angles of from 50°to 70°, the larger angles toward the base, secondaries usually with a double curve, bent downward within 2 to 3 mm. of the midrib, then gently curved apically to just within the margin where they become strongly curved upward and nearly parallel with the margin, often with a strong abaxial branch in the outer ¼ to ⅛; prominent looping and anastomosing of tertiary branches along the margin; tertiary venation a coarse, irregular mesh but with varying, flexuous cross-ties, the meshes from

2 to 4 mm. in dimension; within the tertiary network an even-textured, quadrangular network of fourth order, the meshes averaging 0.5 mm. in dimension; the quartary mesh contains an ultimate areolation of quintary meshes, 0.2 to 0.3 mm. in dimension, within these are one or two, simply branched, tapering, free-ending veinlets; texture firm to heavy; frequency, abundant.

Discussion.—The fine-even-textured ultimate areolation is characteristic of these fossil leaflets. They are among the common species at the Kisinger Lakes locality and occur more rarely at the Boysen site. Many of the leaflets are finely preserved showing clearly the ultimate details of venation.

The fossil leaflets correspond with those of several living species, such as *C. septentrionalis* (*villosa*) Sauer native to tropical Mexico, *C. maritima* Thouars from southern Mexico and Central America, *C. gladiata* (Jacq.) de Candolle from Kiansu, China, *C. brasiliensis* Martius ex Bentham from southwest Mexico and northern South America and *C. acuminata* Rose of southwest Mexico. There are also marked resemblances to the leaflets of several species of *Erythrina;* the finer venation of the leaflets in the two genera is closely similar. However, the common tendency to acrodromy in *Erythrina*, the lack of tertiary branches from the outer extremities of the secondaries, and the expanded petiole serve to distinguish the foliage of this genus from *Canavalia.*

Berry (1916:248, pl. 53, figs. 3–6) described *C. eocenica* from the clay beds of Henry County, Tennessee, the age of which is probably early Middle Eocene. These fossils are similar to the fossil leaflets of *Canavalia* from Wyoming, but they show mucronate rather than acute apices, and a greater development of inter-secondaries. The similarities and differences indicate that we are dealing with the same genus but a different species.

The living species of *Canavalia,* of which about 50 species are recognized, are large, vigorous lianes, growing in the tropics of both hemispheres. The plants are found "trailing on open ground, or climbing on bushes and trees, both as a natural pioneer of dunes, mountain ravines, and river floodplains and as a weedy invader of abandoned fields, roadsides and second growth scrub, usually in lowlands but occasionally above 1000 m. elev." (Sauer, 1964:145, concerning *C. brasiliensis*).

Occurrence.—Kisinger Lakes, Boysen, Ramshorn. Type PA 5646; paratypes PA 5636, PA 5642, PA 5730; topotype PA 5728.

Genus *Carpites* Schimper

Carpites araliodes, new species

(Pl. 15, fig. 4)

Description.—Round, many-valved capsules, width 1 cm., depth 8 mm.; a central columella projects as a point; peduncles 7 mm. long.

These objects are the fruits of some genus of the Araliaceae, probably those of *Dendropanax* (*Gilibertia*) whose fossil leaves occur with the fruits; frequency, common.

Occurrence.—Tipperary, Kisinger Lakes. Type PA 5659; topotypes PA 5660, PA 5661, PA 5662.

Family FAGACEAE

Castaneoides new genus

Castaneoides aequalita, new species

(Pl. 2, fig. 4)

Description.—Leaves long-ovate, length 11–14 cm., width 3–3.5 cm.; base broadly cuneate; apex acute; petiole stout; margin strongly dentate, the dentations spine tipped and defined proximad by relatively deep, narrow sinuses, teeth spaced approximately 1.3 cm.; midrib stout; 16 to 18 pairs of opposite or subopposite secondaries, originating at 45° to 50°, straight, parallel, slightly curved at their extremities, craspedodrome, spaced 2 per cm.; tertiary venation percurrent, nearly perpendicular to the secondaries, spaced 9 per cm.; quartary venation a fine, quadrangular mesh, the opening from 0.4 to 0.5 mm. in dimension; no mesh or higher order present; texture coriaceous; frequency, rare.

Discussion.—In certain instances it appears to be impossible to make a satisfactory generic separation of *Castanopsis, Castanea, Pasania* and certain Asiatic oaks on the basis of foliar characters. The leaves of these genera tend to exhibit the rather closely spaced, parallel or sub-parallel secondaries, percurrent tertiaries and a closed, quadrangular, quartary mesh. The marginal dentations also tend to be similar. Therefore I have used a new generic epithet which indicates relationship of the fossil to this group. The fossil probably represents the foliage of a *Castanea* similar to *C. Seguinii* Dode or a *Castanopsis* similar to *C. fissa* Rehder. The fossils also recall the leaves of such Asiatic species of *Quercus* as *Q. variabilis* Blume.

Castanea seguinii and *Quercus variabilis* are common members of the mixed mesophytic forest formation of China. The evergreen broad-leafed forest formation of China, except in its more tropical aspects, is dominated by a great number of species of evergreen oaks, *Pasania* and *Castanopsis*. Wang mentions 150 species of oaks (1961:136; see also 133–155). In this area there is abundant summer rainfall and the average temperature varies from 60°F to 68°F. Extreme minima over most of the area occupied by this evergreen forest vary from 20°F to 25°F. The southern portions approach a tropical climate, but the evergreen forest as a whole may be considered warm temperate in aspect.

Occurrence.—Tipperary, Kisinger Lakes, Type PA 5604; topotype PA 5605.

Family MELIACEAE
Genus *Cedrela* P. Braun
Cedrela schimperi (Lesquereux), new combination
(Pl. 16, figs. 1, 2)

Juglans schimperi Lesquereux, U.S. Geol. Surv. Terr. Rept., vol. 7, p. 287, figs. 5–10, 1878; Knowlton, U.S. Geol. Surv. Prof. Paper 131, pp. 159–160, 1923; Berry, U.S. Geol. Surv. Prof. Paper 165, p. 68, 1930.
Juglans occidentalis Newberry, U.S. Geol. Surv. Mon. 35, p. 34, pl. 66, figs. 2–4, 1898; Knowlton, U.S. Geol. Surv. Prof. Paper 131, pp. 158–159, 1923; Berry, U.S. Geol. Surv. Prof. Paper 165, p. 68, 1930.
Ficus ungeri Lesquereux, Berry, *ibid.*, p. 70, pl. 14, fig. 5.

These fossils are the impressions of *Cedrela* leaflets and not those of "a fine large walnut" as Berry supposed. Knowlton tried to show that *J. schimperi* was distinct from *J. occidentalis*, but his criteria are not convincing. The relatively long petiolule, the wide angle of the secondaries their near approach to the margin, and the fine, semireticulate tertiary venation mark the relationship of these fossils to the foliage of living *Cedrela*.

Description.—Leaflets long-ovate, length 6–12 cm., width 1.8 to 4 cm., greatest width at the middle of the lamina; petiole 7 to 15 mm. in length; apex narrowed and acute; base cuneate to rounded, usually markedly asymmetric; 15 to 20 pairs of subopposite secondaries arising at angles of 70° to 80° in the lower part of the lamina, the angle decreasing to 50° or 60° distally, secondaries only slightly curved except near the margin where they abruptly turn upward and approach parallelism with the margin, finally merging into the margin, flexuous inter-secondaries usually present, the secondary veins often form a series of fine loops along the margin; tertiary venation essentially reticulate although with scattered, percurrent cross-ties, tertiary meshes 1.5 to 3 mm. in dimension, the general trend of tertiary direction about 70° with the secondaries; no tertiary branches from the ends of the secondaries toward the petiole, the branches all directed distally; quartary mesh quadrangular, meshes 0.8 to 1 mm. in dimension; within the quartary meshes are complexly branched, free-ending veinlets of 5th and 6th order; texture firm; frequency, abundant.

The terms *Juglans schimperi* and *Juglans occidentalis* have been used for a variety of fossil leaves, as can be seen by consulting La Motte's catalogue (1952). It is not within the province of this report to untangle all the assignments, but those noted above are pertinent to the occurrences at Kisinger Lakes and Tipperary, Wyoming.

The fossil leaflets resemble those of several living genera of the Meliaceae; *Cedrela*, *Dysoxylum*, *Guarea*, *Swietenia* and *Walsura*. The long-petiolulate leaflets, only slightly asymmetric, with numerous, sub-parallel secondaries originating at large angles and ascending close to the margin, and the fine, quadrate areolation both serve to distinguish the leaflets of *Cedrela* from the other genera, excepting *Swietenia* (Smith, 1960). It is probably not always possible to separate these two closely related genera on the basis of foliage. There are also resemblances to the leaflets of *Spondias* in the Anacardaceae. *Cedrela* leaflets may be distinguished from those of *Spondias* by their fine, quadrangular, quartary mesh and by the lack of abaxial branching at the ends of the secondaries. *Spondias* leaflets often have a well-marked, marginal vein.

There is sometimes clear evidence, in the case of thin, leaf-bearing strata, that the particular layer was deposited in a comparatively short time, perhaps in one season. At Kilgore, Nebraska, the fossil-bearing layer contains abundant seeds of *Platanus* and the winged fruits of *Cedrela*. This indicates that the fossils were deposited in the fall, when these seeds are abundantly available. At the Wyoming localities no winged fruits of *Cedrela* have yet been found. Considering the abundance of leaflets, this raises some doubt as to their correct identification. Were the beds deposited in spring or early summer?

The fossil leaflets show excellent correspondences with those of the living *C. angustifolia* A. P. de Candolle of tropical Mexico, or those of *C. lilloi* C. de Candolle of Argentina. *C. brunellioides* Rusby from Boliva has dentate leaflets. *C. mexicana* Roemer is a common streamside tree of tropical North America.

Occurrence.—Kisinger Lakes, Tipperary. Hypotypes PA 5663, PA 5664; topotypes PA 5665, PA 5666, PA 5667.

Family CUPRESSACEAE
Genus *Chamaecyparis* Spach
Chamaecyparis sp.
(Pl. 2, fig. 1)

Discussion.—A small fragment of the characteristic foliage is the only trace of this species found in the floras. In the absence of cones it is not possible to assign the fragment to any particular species. It is probably the same species as that found at Florissant (MacGinitie, 1953:89). The rarity of this form suggests that it grew at some upland site and was rafted down the river which deposited the Kisinger Lakes flora.

Occurrence.—Kisinger Lakes. Type PA 5602.

INCERTAE SEDIS
Cf. *Cissus marginata* (Lesquereux) Brown
(Pl. 14, fig. 1)

Cissus marginata (Lesquereux) Brown, U.S. Geol. Surv. Prof. Paper 375, pp. 79–81, pls. 53, 54, 1962.

This fragmentary leaf is illustrated to show the remarkable resemblance to the Late Cretaceous and Paleocene species discussed by Brown. There can be little doubt that it represents a species of this extinct genus, whatever its relationship may be. I doubt that the assignment to *Cissus* is tenable, there are too many points of difference. The scalloped margin and the strong side primaries decurrent into the petiole are not found in the foliage of *Cissus;* neither are the various degrees of decurrency of the lamina along the top of the petiole (perfoliate petiole). The venation is that of *Platanus;* the characteristic, even-textured quadrangular, fourth order net is present and the ultimate venation is also characteristic of *Platanus.* The indicated trifoliate arrangement of the leaves or leaflets does not necessarily contradict an assignment to the Platanaceae, and so in spite of the trifoliate nature these leaves probably represent some extinct platanoid genus. Frequency, rare.

Occurrence.—Tipperary. Hypotype PA 5651.

Family CORNACEAE
Genus *Cornus* Linné
Cornus sp.
(Pl. 24, fig. 3)

Discussion.—The single incomplete leaf figured clearly exhibits the character of *Cornus* leaves, but I do not consider it good practice to found a species on a single, incomplete specimen. It appears impossible to separate the "flowering" dogwoods (*C. florida, C. nuttalli*) from the osier type (*C. stolonifera*, etc.) on the basis of foliage. Since, after extensive collecting, only one leaf impression has been found, it may be that the *Cornus* of the Kisinger Lakes area in the Eocene lived at some distance from the site of deposition, perhaps on some higher slope.

Many species of fossil *Cornus* have been described from the western states (see La Motte, 1952:132–133), most of them from the Paleocene. It is probable that this represents considerable over-speciation. Since the foliage of the existing and fossil dogwoods shows so much similarity among the species, it seems unprofitable and misleading to indicate likenesses with any living or fossil species. This species is clearly derived from a species found in the underlying Wind River Formation which will be described in a later publication.

Occurrence.—Tipperary. Type PA 5695.

Family ARALIACEAE
Genus *Dendropanax* Decaisne and Planchon
Dendropanax latens, new species
(Pl. 13, fig. 1; pl. 19, figs. 1, 2; pl. 29, fig. 1)

Description.—Leaf ovate, greatest width just below the middle of the lamina, rarely trilobed, with elongated lobes arising near the middle of the lamina and extending to nearly the apex of the middle lobe; length 6–9 cm., width 3–4 cm.; apex acuminate; base cuneate, margin entire; petiole relatively heavy, elongate, 3–4.5 cm. in length, midrib slender, straight; secondaries 3 to 4 pairs, irregularly spaced, the basal pair opposite, pseudo-palmate, arising 2 to 4 mm. above the top of the petiole at angles of 25° to 30°, the succeeding secondaries at angles of 30° to 35°, often a pair of slender secondaries, marginal or nearly so, in the lower half of the lamina, secondaries pursuing a relatively straight course to just within the margin, giving off a series of tertiary branches on the outer (obmedial) side of the secondary in its upper half, these tertiary branches forming a series of diminishing loops elongated parallel to the secondaries, the basal secondaries and the extremities of the succeeding secondaries together with the tertiary loops forming a prominent marginal vein, the margin thickened and slightly rolled; inter-secondaries at slightly larger angles than the true secondaries, branching about 1 cm. within the margin and forming connecting tertiaries between the secondaries, tertiary venation tending to stream across the lamina at approximately 60° with the midrib, forming a coarse, irregular mesh, angular-polygonal; quartary veinlets forming a markedly irregular, polygonal mesh within the tertiary mesh, the openings 1 to 1.5 mm. in size; finer venation obscure; texture firm; frequency, common.

Discussion.—These fossil leaves exhibit all the characters found in the foliage of the existing genus *Dendropanax* (*Gilibertia*). A few of the fossils show the trilobed shape also found among living species. The most similar living species is *Dendropanax arboreus* (Linné) Decaisne and Planchon of southern Mexico and Central America. The genus is, at present, confined to tropical or subtropical regions; about 20 species have been recognized in the humid to semi-deciduous tropical forests from southern Mexico and the West Indies southward into South America and in southern Asia. *D. arboreus* species is common in the bosque tropical deciduo (moister phase) or the bosque tropical subdeciduo (Rzedowski and McVaugh, 1966:15, 23) along the southwest coast of Mexico. It is a medium to large tree growing with *Guarea, Thouinidium, Ficus, Brosimum, Enterolobium, Cedrela, Luehea, Persea* and the like. The species also extends into the moister, mesic montane forests in company with *Liquidambar, Carpinus, Quercus, Ilex, Meliosma, Persea, Tilia* and other "temperate" types. *Dendropanax* is a member of the evergreen, broad-leaved forest of southern China and Taiwan where it may be subject to light frosts.

Dilcher and Dolph (1970) have identified leaves of *Dendropanax* from the Claiborne (early (Middle Eocene) deposits of western Tennessee. Their species, *D. eocenensis*, differs from *D. latens* in the broader leaves, some of which show five lobes. The authors also suggest that the Claiborne species may be conspecific with a Paleocene form described by Lesquereux from the Fort Union of southwestern North Dakota, and called *Aralia acerifolia* (now *A. dakotana*). There are *Dendropanax* leaves in the late Early Eocene Wind River flora, which occurs just below the Kisinger Lakes early Middle Eocene flora.

Occurrence.—Tipperary, Kisinger Lakes. Type PA 5716; paratypes PA 5679, PA 5680, PA 5647.

Family EBENACEAE
Genus *Diospyros* Linné
Diospyros mira Berry

Diospyros mira Berry, U.S. Geol. Surv. Prof. Paper 165, p. 76, pl. 14, fig. 7, 1930.

The shape and the secondary and tertiary venation of this fossil leaf correspond well with the characters of *Diospyros virginiana* Linné. However, without more material and the presence of well-preserved areolation, the identification must remain somewhat doubtful.

Occurrence.—Tipperary.

Family ACERACEAE
Genus *Dipteronia* Oliver
Dipteronia wyomingense (Berry), new combination
(Pl. 25, fig. 4; pl. 34, fig. 1)

Dryophyllum wyomingense Berry, U.S. Geol. Surv. Prof. Paper 165, p. 65, pl. 8, figs. 6–9, 1930;
Myrica ludwigii Schimper, Berry, *ibid.*, p. 69.

The fossils described by Berry cannot be Fagaceous; the margin is doubly serrate and the tertiaries are more nearly perpendicular to the midrib than to the secondaries. The bifurcating of the secondaries near the margin is not typical of Fagaceous leaves. It is clear that the fossils do not possess characters typical of *Quercus, "Dryophyllum," Castanopsis, Castanea* or *Lithocarpus*. There is a certain similarity to the leaves of the living *Chaetoptelea* or *Ulmus*, but the finer details of venation are greatly different from those in the foliage of any living genera of the Ulmaceae. Recent collections prove that the fossil leaflets are those of *Dipteronia*. An incomplete leaf is illustrated.

Description.—Leaves odd-pinnately compound; estimated length 24–35 cm.; leaflets opposite, long ovate, length 5–12 cm., width 2.3 to 3.5 cm., base asymmetric, broadly cuneate to rounded, apex extended-acute, margins coarsely serrate, often simple serrate but occasionally remotely double serrate; midrib strong, about 15 pairs of subopposite to alternate secondaries originating at 75° near the base, the angles decreasing to 45° toward the apex, essentially craspedodrome but often bifurcating near the margin and sending a branch to the apex of the sinus, this branch met by another from the next succeeding secondary, the two tertiaries uniting in a gland at the vertex of the sinus; tertiary venation irregularly percurrent, more rarely reticulate; quartary venation a relatively coarse, reticulate, polygonal mesh; quintary venation consisting of complexly branched, free-ending nervilles; texture firm; frequency, common.

The peculiar tertiary branches leading to the sinuses and the branched, free-ending quintaries confirm the assignment of these fossil to *Dipteronia*. This appears to be the earliest Tertiary record of the genus. Well-preserved fruits occur in the Republic flora of eastern Washington, of latest Eocene age. Both foliage and fruits are found in the Florissant flora of early Oligocene age (MacGinitie, 1953:142). The foliage of the Florissant species is about half the size of that from the Kisinger Lakes Quadrangle. The latest occurrence of *Dipteronia* in the American Tertiary appears to be that at Bridge Creek, Oregon, of later Oligocene age. According to our present knowledge, the genus seems to disappear before the beginning of the Miocene.

There is probably only one living species of the genus *Dipteronia, D. sinensis* Oliver of the deciduous, broad-leafed forest formation of the southern parts of Shensi and Kansu and adjacent Honan, and also in Hupeh and Szechuan. Some taxonomists recognize two species but this has not been proved to be correct. *D. sinensis* grows in temperate habitats and is moderately frost-hardy; it is in no sense of tropical affinities.

The fossil leaflets differ from those of the living species in the following characters: (1) less prominent double serrations, (2) secondaries more curved, (3) slightly more percurrency of the tertiaries. These are relatively minor differences, but show slight evolutionary changes since the Eocene. The venation of *Dipteronia* shows a striking resemblance to that of several species of *Acer; A. circinatum* Pursh for example.

Occurrence.—Kisinger Lakes, Tipperary. Hypotype PA 5699; topotypes PA 5703, PA 5704.

Family EQUISETACEAE

Genus *Equisetum* Linné

Equisetum tipperarense Berry

Equisetum tipperarense Berry, U.S. Geol. Surv. Prof. Paper 165, p. 61, pl. 6, figs. 4–10, 1930; Univ. Calif. Publ. Geol. Sciences, vol. 83, p. 90.

This large horsetail has left abundant remains in the Green River deposits and in the Wind River basin. Stem fragments of 3 to 4 cm. in diameter are not uncommon, indicating that the height reached by the plants must have been from 10 to 15 feet. It is probable that *E. winchesteri* Brown from the Green River formation of northern Colorado (MacGinitie, 1969:90) is conspecific with *E. tipperarense* Berry, but the fossil material, which consists of stem fragments, does not seem to warrant synonymizing the two species. The fossil species is similar to the living *E. giganteum* Linné of South America.

Occurrence.—Kisinger Lakes, Tipperary.

Family MYRTACEAE

Genus *Eugenia* Linné

Eugenia americana (Knowlton) MacGinitie

(Pl. 24, figs. 1, 2)

Eugenia americana (Knowlton) MacGinitie, Univ. Calif. Publ. Geol. Sciences, vol. 83, p. 122, 1969.

This species is found in the upper horizon at Kisinger Lakes in company with *Salix cockerellii* and *Platanus brownii* (or *wyomingensis*). This is a common association in the upper Parachute member of the Green River Formation in northwestern Colorado. This indicates a transition toward a Green River flora in the later Aycross beds. The *Eugenia* from Kisinger Lakes cannot be satisfactorily separated from the Green River species, although there may be varietal differences. This Middle Eocene species differs from that at Florissant, Colorado (MacGinitie, 1953:153), in its more slender, elongate form and the more numerous secondaries. Only a few impressions were found at the Kisinger Lakes locality.

Occurrence.—Kisinger Lakes. Hypotype PA 5694.

Family TAXODIACEAE

Genus *Glyptostrobus* Endlicher

Glyptostrobus europaeus (Brongniart) Heer

Glyptostrobus europaeus (Brongniart) Heer, Flora Tertiaria Helvetiae, vol. 1, p. 51, pl. 19, pl. 20, fig. 1, 1855; Berry, U.S. Geol. Surv., Prof. Paper 92, 46–48, 1924; *ibid.* 144, pl. 49, figs. 1, 2, Prof. Paper 156, p. 52, pl. 7, figs. 7–9, 1930.

The impression of a spray of *Glyptostrobus* bearing a few staminate aments was found at Tipperary, associated with the impressions of *Equisetum* and swamp monocots. The references to Berry's work are given to show the occurrence of the species in the Eocene of the Gulf States. It is evident, from the numerous citations, that the genus was widespread in the Tertiary of the Northern Hemisphere, and particularly in the Paleogene. *Glyptostrobus* has been identified from scores of localities in North America, Europe and Asia. Some of the foliage assigned to *Glyptostrobus* may be that of *Taxodium*, but remains of the unique cones confirm the identification.

The living *Glyptostrobus* comprises small trees inhabiting low ground and swampy areas in the evergreen, broad-leafed forest formation of southeastern China. Individual trees are often scattered along the borders of rice fields. It is not clear from the literature on the genus that the species has been found in the wild state.

Occurrence.—Tipperary. Topotype PA 5687.

Family AQUIFOLIACEAE

Genus *Ilex* Linné

Ilex sclera, new species

(Pl. 23, figs. 1–4; pl. 33, fig. 4)

Description.—Leaves broadly ovate to orbicular, length 3.4 to 6 cm., width 2 to 2.6 cm.; apex varying from bluntly pointed to emarginate; base cuneate, decurrent along the upper petiole; petiole stout 0.7 to 1 cm. in length; margin entire; midrib slender; 5 to 6 pairs of subopposite secondaries, irregularly spaced, arising at 55° to 60°, complexly branched in the outer one-third to one-half of the lamina and forming a series of random loops along the margin, a notable character of the fossil leaves are these complex, angular, polygonal loops which may occupy as much as ½ of the area of the leaf; tertiary venation a coarse, irregular, reticulate mesh, the openings 1.5 to 3 mm. in greatest dimension; within the tertiary mesh is a much finer mesh of quartary veinlets, aligned roughly perpendicular to the secondaries, the meshes 0.2 to 0.3 mm. in dimension; further details of venation obscure; texture coriaceous; frequency, common.

Discussion.—These fossil leaves with their short heavy petiole, orbicular shape and strange, irregular venation, represent the foliage of an *Ilex* similar to the leaves of the existing *I. rotunda* Thunberg of southeast Asia (Korea, Japan, Taiwan, southern China, Indo-China) and *I. theezans* Martius of Brazil. Although several species of *Ilex* flourish in the north temperate climates of western Europe and America, for example, *I. aquifolium* Linné of England and *I. decidua* Walter, and *I. glabra* of the states east of the great Plains, and are adapted to severe winters, the genus as a whole is tropical in its distribution. *Ilex* is found worldwide, from Australia to Africa, and Europe to America. Li (1963:451–464) lists 23 species in Taiwan alone and there are from 150 to 180 species recognized over the world. The fossil species closely resembles the leaves of *I. rotunda* which grows in warm temperate to tropical moist climates.

Occurrence.—Kisinger Lakes, Tipperary. Type PA 5691; paratypes PA 5689, PA 5690; topotypes PA 5692, PA 5693.

Family JUGLANDACEAE

Genus *Juglans* Linné

Juglans alkalina Lesquereux

Juglans alkalina Lesquereux, U.S. Geol. Surv. Rept. Terr., vol. 7, p. 288, pl. 62, figs. 6–9, 1878; Berry, U.S. Geol. Surv. Prof. Paper 165, p. 68, pl. 10, figs. 1, 2.

This species is of doubtful status. In spite of Berry's figures, leaflets assignable to *Juglans* are uncommon and poorly preserved. For the present it is kept as a member of the flora, with the proviso that the species is questionably present. Pollen grains of *Juglans* have been found in the rich and well-preserved pollen flora, but they are relatively rare. Considering the enormous production of pollen by trees of this genus, they must have been rare in the flora or at some distance from the site of deposition.

Occurrence.—Tipperary (?).

Family LAURACEAE

Genus *Laurophyllum* Goeppert

Laurophyllum fremontensis (Berry), new combination

(Pl. 6, figs. 2, 3; pl. 8, fig. 2; pl. 18, fig. 1)

Laurus fremontensis Berry, U.S. Geol. Surv. Prof. Paper 165, p. 74, pl. 13, fig. 3, 1930.

Description.—Leaf long-ovate; base cuneate; apex acute or acuminate; margin entire; petiole relatively slender, entire length not preserved; midrib straight, slender; 10–12 pairs of chiefly opposite secondaries, originating at 70° to 80° in the central part of the leaf but curved upward within 5 mm. of the midrib, the main course of the nerve at about 35° to 40° with the midrib, secondaries approaching within 1 mm. or less of the margin, following up along the margin, with many, fine tertiary ties to the succeeding secondary, finally merging with a distinct marginal vein; tertiary venation essentially percurrent, the tertiaries perpendicular to the secondaries, either entire or once branched half way between the secondaries forming Y's, 4–6 tertiaries per cm. of secondary; quartary venation an angular network, the nervilles approximately perpendicular to the tertiaries, a double row of quadrangular or pentagonal areoles between each pair of tertiaries, the areoles approximately 1 mm. in average dimension; quintary venation, a fine, quadrangular, closed mesh, the areoles 0.2 to 0.3 mm. in dimension; texture firm to coriaceous; frequency, not common.

Discussion.—It appears impossible to distinguish the leaves of certain species of *Phoebe* and *Machilus* on the basis of venation. The fossil leaves closely resemble those of *Machilus velutina* Champion from the mixed evergreen forest of southern China. However, they are also much like the leaves of *Phoebe sheareri* Hemsley from the same general area. Thus the name *Laurophyllum* is used rather than implying more than is justified in the identification. Both *Phoebe* and *Machilus* are common genera in what Wang (1961:129) has called the evergreen, broad-leaved forest formation. This vegetational formation occurs in the southeastern provinces of China and in neighboring Hainan and Taiwan in a warm, humid, subtropical climate.

Berry illustrated this species but described (p. 74) something else that does not fit the illustration, probably confusing this with his "Ficus."

Occurrence.—Kisinger Lakes, Tipperary. Hypotypes PA 5615, PA 5627, PA 5672.

Genus *Laurophyllum* Goeppert
Laurophyllum quotidiana, new species
(Pl. 6, fig. 1; pl. 12, fig. 2)

Description.—Leaf ovate, length 7 to 9 cm., width greatest in the upper half, 2.5 to 4 cm.; apex rounded to acute; midrib slender; 6 to 7 pairs of opposite to subopposite secondaries, unevenly placed, tending to be exarcuate (curved upward) at their juncture with the midrib, arising at 45°, gently curved upward to within 3 to 4 mm. within the margin, where they curve upward and form a series of prominent loops with tertiary branches from the succeeding secondary, the outer extension of the loops 1 to 3 cm. within the margin, secondaries rarely bifurcating at half their length; tertiary venation largely percurrent, the veins perpendicular to the secondaries, spaced 4 to 6 mm. apart; quartary venation a quadrangular mesh, the opening from 1 to 2 mm. in greatest dimension; an ultimate, fine, closed mesh of quintary veinlets; texture heavy. Frequency, rare.

Discussion.—These leaves have the unmistakable venation of certain genera of the Lauraceae. In what might be called the *Laurus* type of venation the ultimate mesh consists of fine, quadrangular, closed areoles, the ultimate mesh usually meaning the fifth order or quintary mesh for these leaves. Among lauraceous genera sharing this type of venation are *Cryptocarya, Litsea, Phoebe, Persea* and *Machilus.* Distinctions may sometimes be made among these genera on the basis of leaf shape, number and angle of secondaries, and behavior of the tertiaries. However, it is my opinion that, in some cases, the leaves of *Cryptocarya, Persea,* and *Phoebe,* for example, cannot be separated on the basis of venation. The fossil leaves considered here might belong to any one of those three genera, and it seems impossible to reach a decision as to which one. They appear close to the foliage of certain species of *Persea,* such as *P. cinerascens* Blake of southern Mexico, and thus I have chosen this species as *tentatively* the living representative of the fossil species. *Machilus* and *Persea* have both been considered as *Persea* (Kostermans, 1962). The ecological requirements of the living genera mentioned are similar: along river courses in the deciduous or semi-deciduous forests, or members of the evergreen, broad-leafed forests, in tropical climates.

Occurrence.—Kisinger Lakes, Tipperary. Type PA 5616; paratype PA 5641.

Family LEGUMINOSAE
Genus *Leguminosites* Bowerbank
Leguminosites inlustris, new species
(Pl. 11, figs. 2, 4)

Description.—Leaflets ovate, length 3.5 to 5 cm., width 1.5 to 2.5 cm.; base rounded, apex blunt or apiculate; margin entire; petiolule short, of medium weight; midrib slender straight; 6 to 7 pairs of secondaries, curved upward (exarcuate) and simply looped within the margin, tending to branch 5 mm. Within the margin, tertiaries reticulate to semi-percurrent, irregularly distributed, uneven as to course, the tertiary meshes 3 to 4 mm. in dimension; quartary venation a fine, even-textured mesh, the opening 0.3 to 0.4 mm. in dimension; finer venation apparently consisting of free ending simply branched veinlets with rounded thickenings at the junctures, giving the leaf a dotted appearance; texture firm; frequency, not common.

Discussion.—These leaflets resemble the foliage of several genera of tropical legumes— *Dioclea* and *Inga* for example. The foliage of both these genera show the punctate dots. This last character is also shared by the foliage of *Shepherdia* and *Hymenocardia*, among others. The closest correspondences appear to be with the foliage of *Dioclea reflexa* Hooker of the American tropics, a forest tree of the moist forests from southern Mexico to Argentina.

Occurrence.—Kisinger Lakes. Type PA 5638.

Leguminosites mira, new species
(Pl. 8, fig. 3; pl. 10, figs. 1, 2; pl. 11, fig. 3)

Leaflet ovate, length 3.5 to 11 cm., width 1.7 to 4.4 cm., 1/w ratio 2.1; apex acuminate, narrowed to a rounded point; base broadly cuneate; petiolule stout, 2 to 11 mm. in length, marked by cross-ridges; margin entire; midrib relatively slender, tapered, strongly broadened as the petiolule is approached; 7 to 12 secondaries, arising at 25° to 30°, strongly bowed outward (exarcuate), then bowed upward near the margin to form prominent loops with the succeeding secondary, secondaries continued to form a looping series within the margin; tertiaries markedly reticulate, a few exarcuate cross-ties; scattered, strong quartaries form a coarse, polygonal network, the mesh opening averaging 3 mm. in dimension; within the quartary mesh quintary veinlets form a complex angular, reticulate mesh whose openings are approximately 0.7 mm.; ultimate details of venation obscure; texture firm; frequency, common.

Discussion.—These leaflets are marked by the extremely complex angular areolation and the pronounced marginal looping of the secondaries. They resemble the leaflets of several existing tropical legumes, particularly those of certain species of *Lonchocarpus* and *Derris*. There are also close resemblances to the leaflets of *Andira*, *Piscidia* and *Dioclea*. However, *Lonchocarpus* leaflets appear to furnish the best correspondences, especially in the lack of well-marked intersecondaries and in the characteristic marginal looping of the secondaries. Considering the difficulty or impossibility of assigning fossil legume leaflets to any one genus, especially in a flora of this age, the best interpretation of these fossils seems to be that they represent the large leaflets of some tropical legume.

These fossil leaflets show certain similarities to *Lonchocarpus coriaceus* Potbury from the Oligocene La Porte flora of the northern Sierra Nevada but differ in the number and angle of the secondaries and in the finer tertiary net.

Occurrence.—Kisinger Lakes, Tipperary. Type PA 5634; paratypes PA 5633, PA 5626, PA 5640.

Leguminosites occulta, new species
(Pl. 18, figs. 3, 4)

Description.—Leaf odd-pinnate, five leaflets preserved; terminal leaflet ovate; length 3.5 cm., width 1 cm.; base long-cuneate; apex acute, mucronate, petiole stout, 3 mm. in length; margin entire; lateral leaflets ovate; length 2.5 cm., width 1 cm., nearly sessile, petiolule 1 mm. in length; base cuneate; apex acute or rounded-mucronate; margin entire; midrib stout; 5–7 pairs of subopposite secondaries arising at 45°, pursuing a straight course and looping just within the margin; tertiary venation coarsely reticulate, with rare cross-ties; quartary venation quadrangular reticulate the meshes from 0.4 to 0.5 mm. in dimension; quintary venation a fine even textured network, the meshes 0.1 to 0.15 in dimension; texture firm; frequency, rare.

Discussion.—It does not seem possible to make a detailed identification with any modern genus. The fossil could represent a *Cassia*, but there are at least a score of other genera with similar foliage. Thus any assignment to a living genus would be assuming unjustified knowledge.

Occurrence.—Kisinger Lakes. Type PA 5675; paratype PA 5674; topotype 5676.

Leguminosites wyomingensis (Berry), new combination
(Pl. 10, fig. 3; pl. 12, fig. 4)

Fagara wyomingensis Berry, U.S. Geol. Surv. Prof. Paper 165, p. 71, pl. 11, fig. 4.

Description.—Leaflets ovate; Length 4.6 cm., width 2.5 to 3 cm.; apex acute; base cuneate to rounded; petiolule slender, 3–4 mm. in length; margin entire, but with a tendency to ruffling; midrib slender; 5 to 6 secondaries arising at angles of 60° to 70°, the larger angles near the base, secondaries curving upward along the margin, approaching the margin within 1 to 2 mm., complexly looped with the succeeding secondary; tertiary venation mainly reticulate, although with scattered cross-ties, tertiary meshes irregular, 3 to 4 mm. in greatest dimension, occasional tertiary branches curve back toward the midrib and branch; quartary venation reticulate, the meshes 0.5 to 0.7 mm. in width; within the quartary mesh is a fine, complex, roughly quadrangular areolation, the meshes of two orders, the finer less than 0.1 mm. in size; texture firm; frequency, not common.

Among living plants the fossil leaflets most closely resemble those of several species of *Derris*, and *Lonchocarpus*. These two genera appear closely related. The closest correspondence appears to be with leaflets of *Derris trifoliata* Loueiro of Kwantung province, China or *D. negrensis* Bentham of northern South America. There are also significant resemblances to the leaflets of *Lonchocarpus sericeus* (Poir.) Standley, common in the tropics from Sinaloa, Mexico to northern South America. The living species of *Derris* (50) and *Lonchocarpus* (120) are widely distributed in the tropics of both hemispheres.

Occurrence.—Kisinger Lakes, Tipperary. Type PA 5635; paratype 5644; topotype PA 5645.

Family HAMAMELIDACEAE
Genus *Liquidambar* Linné
Liquidambar callarche Cockerell

Liquidambar callarche Cockrell, U.S. Nat. Mus. Proc., vol. 66, p. 8, pl. 1, fig. 6, pl. 2, fig. 5, 1925.

This species is rare in both the Green River beds and the Aycross Formation, although pollen of the genus is relatively common. Only two incomplete leaves have been found at the Tipperary locality.

Occurrence.—Tipperary. Topotypes PA 5710, PA 5711.

Family TILIACEAE
Genus *Luehea* Willdenow
Luehea newberryana (Knowlton), new combination
Pl. 13, fig. 3; pl. 27, fig. 1; pl. 28, figs. 3, 4; pl. 34, fig. 2)

Carpites newberryana Knowlton, U.S. Geol. Surv. Prof. Paper 131, p. 174, 1923; Berry, U.S. Geol. Surv. Prof. Paper 165, p. 77, pl. 14, figs. 2, 6 (?), 1930; MacGinitie, Univ. Calif. Publ. Geol. Sciences, vol. 63, p. 126, pl. 31, figs. 1, 4, 1969.

Description.—Five-valved, loculicidal capsules, arising from short (3 to 5 mm. in length), peduncles. The capsule originates from a pronounced ring-shaped enlargement, which bore the perianth segments. Carpels ovate with acute apices, the bases rounded; size variable, from 1 to 2.3 cm. long; the surface with a feltlike texture but not ornamented.

Description.—Leaves ovate, length 9 to 12 cm., width 4.6 to 5 cm.; petiole slender, 2.5 to 3 cm. in length; margin doubly serrate in the upper ¾ of the lamina, the serrations 7 to 8 per cm. of margin, the teeth apically oriented and the sinuses arcuate; apices acute to extended; base cuneate; venation acrodrome, a pair of prominent side-primaries extending to the upper third of the leaf, arising at 30°, forming a smooth, gentle curve, approaching close to the margin and finally entering a pronounced marginal dentation; another slender, short pair of acrodrome veins may be present, arising at angles of 60° and extending for 2 cm. along the margin, finally sending a branch to a small dentation; 2 to 3 pairs of secondaries in the upper half of the leaf, arising at angles of 35°, curved upward just within the margin and sending a branch to a dentation; five curving secondaries on the marginal side of the main acrodrome veins forming pronounced loops and with tertiary branches to the marginal teeth; tertiary venation markedly percurrent, the veins approximately perpendicular to the midrib, running across the leaf and gently curved adaxially (down) near the margin, the tertiaries spaced 6 to 7 per cm.; quartary venation of thin cross-ties perpendicular to the tertiaries, often bifurcated; quintary venation a fine, even, polygonal mesh, the areoles averaging 0.1 mm. in dimension; texture firm; frequency, leaf impressions rare, capsules not rare.

The capsules from the Kisinger Lakes and Tipperary localities are often twice the size of those found in the Green River beds, but otherwise conform in all their characters. Berry (1930b: 77) did not follow through on the statement, "Their botanical relationship remains questionable, although they suggest to me comparisons with the genus *Luehea* Willdenow, of the family Tiliaceae." There can be no doubt that these fossils represent the fruits of *Luehea,* the correspondences are exact as can be seen among the numerous specimens available from the Wyoming sites.

Only a few leaf impressions have been found and these are all imperfect; the plant appears to have grown at some distance from the site of deposition. The characters of the leaves: the thin and almost elegant primary and secondary venation, the numerous percurrent tertiaries aligned perpendicular to the midrib, the thin quartaries aligned parallel to the midrib, and the peculiar, exceptionally fine, closed areolation, prove the affinity of the fossil leaves with those of several genera of the Tiliaceae. They most closely resemble the leaves of the existing *Luehea divarcata* Martius or *L. speciosa* Willdenow of tropical America. There are about 18 species of the genus growing from Argentina to southern Mexico and Cuba. Some of the species are found on mountain slopes at altitudes of 3000 to 4000 feet, while others are of tropical lowland habitat, common along water courses. The genus is essentially tropical in its modern distribution.

There are pollen grains in the Kisinger locality sediments which may be assigned to this genus. A closely similar fossil leaf occurs in the lower part of the Healy Creek Formation (Oligocene), Nenana coal field, Alaska (Personal communication, Jack Wolfe, May 10, 1969).

Occurrence.—Kisinger Lakes, Tipperary. Hypotypes PA 5706, PA 5714, PA 5715, PA 5650.

Family SCHIZAEACEAE
Genus *Lygodium* Swartz
Lygodium kaulfussi Heer

Lygodium kaulfussi Heer, Berry, U.S. Geol. Surv. Prof. Paper 165, pl. 7, figs. 2, 3, 1930.

Impressions of *Lygodium* leaves and sporangia are abundant at both localities. The leaves are somewhat variable as to size and shape, the 4- to 5-lobed forms being the most common. The species is abundant in the Middle Eocene floras of the central Rockies; it is one of the dominants at the Kisinger Lakes, East Fork, Boysen, and Valley localities in Wyoming. The genus comprises about 25 living species of scandent ferns, largely tropical in distribution but extending into the moist, warm-temperate areas of the southeastern United States, Japan and New Zealand. Berry has given a rather full discussion of the fossil species (1924:40). *L. kaulfussi* (Brown preferred the name *Lygodium neuropteroides* Lesquereux) is one of the few species found in the Middle Eocene of both the central Rockies and the Gulf Coast. It is common at the Schoening, Wyoming locality of early Eocene age. Knowlton (1899:672) recorded the species from the "Paleocene" of Yellowstone National Park, but this was probably an error in stratigraphy. The species does not seem to predate the Early Eocene, although *L. coloradense,* a closely similar species, occurs in the Coalmont Formation of North Park, Colorado (Paleocene-Eocene). The earliest record of *Lygodium* in North America is that of *L. pumilum* Brown (1943:141; 1962:45) from the Montana group near Casper, Wyoming.

Occurrence.—All localities. Topotype 5610.

Family PLATANACEAE
Genus *Platanus* Linné
Platanus browni (Berry), new combination
(Pl. 15, fig. 1; pl. 16, figs. 3, 4; pl. 26, fig. 3)

Aralia browni Berry, U.S. Geol. Surv. Prof. Paper 165, p. 75, pl. 13, fig. 5, 1930.

Berry designated this fossil as *Aralia,* using the term in a "general sense" and suggested that such leaves probably represent the genus *Oreopanax.* The venation and outline have little in common with the leaves of *Orepanax;* the characters clearly indicate that the fossils represent the genus *Platanus* and not any member of the Araliaceae. The same error was made in the case of *Platanus wyomingensis* (Knowlton and Cockerell) MacGinitie (1969:104–105). Berry noted the resemblance to *Platanophyllum angustiloba* (Lesquereux) MacGinitie from the early Eocene Chalk Bluffs flora of California. The two do not represent the same species since the California fossils are larger, often with 7 to 9 lobes, and with consistently longer and narrower lobes, and, where the margin is dentate, with camptodrome secondaries.

Berry described the major features of these leaves but omitted the characteristics of the finer venation. The secondaries are predominantly camptodrome; where dentations occur the secondaries branch just within the margin, sending one tertiary branch to the dentation while the weaker tertiary follows up the margin to loop with a branch with the next succeeding secondary. The striking feature of the venation is the fine, guadrangular, quartary net, whose meshes are from 0.3 to 0.5 mm. in width. This network is immersed in a thin, coarse, tertiary network, whose veins are little larger than the quartary veins. The secondaries are markedly curved as they approach the margin, and they tend to be somewhat irregular as to course and spacing.

Berry states, with reference to *P. browni*, "It differs from the Green River *Aralia wyomingensis* Knowlton and Cockerell in its narrower lobes, much deeper sinuses, and much less prominent marginal teeth." *P. wyomingensis* differs from *P. browni* in the greater tendency to craspedodromy of the secondaries and in its coarser quartary mesh. The openings of the quartary areolation in *P. browni* vary between 0.3 to 0.5 mm. in width, while those of *P. wyomingensis* vary from 0.7 to 1.0 mm. in width. The coarser quartary mesh is a distinguishing character of the Green River species and may be correlated with the indicated subhumid environment of this flora with its evidence of a hot, dry season (MacGinitie, 1969:50). All the living species of *Platanus,* which are native to humid summer climates, which I have examined, show the extremely fine quartary mesh, and the same is true of the fossil species of *Platanus* and *Platanophyllum* occurring in fossil floras with indications of a humid climate. The coarser areolation is found in the living *P. racemosa* and its variety *wrightii,* which grow in the subhumid climates of California and Arizona. *P. orientalis* L. (one of two or three living species recognized in the old world), which grows from Greece eastward though the Near East to northwestern India, also shows the coarse areolation in occasional leaf specimens.

Although *P. browni* and *P. wyomingensis* have a rather close resemblance, it appears best to designate them as spearate species on the basis of the difference in areolation, and in the greater tendency in *P. browni* to narrower lobes and deeper sinuses. The marginal teeth, notwithstanding Berry's statement, are much the same in the two forms. Leaves with entire margins are fairly common among the *P. browni* fossils but are extremely rare among the *P. wyomingensis* fossil leaves of the Green River Formation.

It appears from the variety of leaf types among the late Cretaceous and Paleogene fossils referable to *Platanus* that the genus was undergoing rather complex genetic changes. The Eocene, in particular, seems to have been a time of accelerated speciation. However the chaotic state of the taxonomy in *Platanus* obscures many significant relations. Over-speciation and wrong identifications have so confused relationships that a thoroughgoing revision of the genus must be made before much progress can be attained in tracing genetic lineages. Such a revision is beyond the scope of the present paper. Brown (1962:64–65) has given synonymies for the Paleocene *Platanus raynoldsii* Newberry and *Platanus nobilis* Newberry. These may seem too inclusive, but my studies on the Eocene species tend to confirm Brown's ideas. Brown makes a pertinent statement, "However, after observing the great variation shown by leaves from living species of *Platanus,* I am amazed that paleobotanists have had the temerity to describe as many species as they have" (1962:65).

The *"Platanophyllum"* type of leaf with its extremely fine quartary mesh, and 5 to 9 entire lobes, is one of the characteristic plant fossils of the early Eocene in the Rockies, although this form, mainly with 5 lobes, extends to the Late Eocene. It is relatively abundant in the Republic, Washington flora. The living *P. lindeniana* Martius and Galeotti of Mexico may be a derivative of this lineage. *P. rileyi* Ball from the Fayette sandstone (Late Eocene) of Grimes County, Texas, is probably an offshoot of the *"Platanophyllum" whitneyi* complex. *P. rileyi* and *P. wyomingensis* are somewhat similar as to outline but *P. wyomingensis* does not possess the appendiculate, basal lobe. The finer details of venation in *P. rileyi* are not preserved, and thus its place in a lineage may never be decided. *P. browni* may be another offshoot of the *"Platanophyllum"* complex, since it shows some of the typical characteristics of the group. Leaves of *Platanus* are among the most abundant fossils at both localities in the western Wind River basin.

Occurrence.—Kisinger Lakes, Tipperary, Ramshorn. Hypotypes PA 5654, PA 5668, PA 5669.

Platanus intermedia, new species
(Pl. 17, fig. 2)

Description.—Leaf ovate, length 18 cm., width 17 cm.; petiole not complete, length 3+ cms.; leaf shallowly 3-lobed, the side lobes projecting 3 cm. beyond the sinus; apex acute; base decurrent along the petiole for 1.5 cm.; margin dentate, with wide, shallow teeth spaced about 1 cm., the inter-dentate sinuses curved, or scalloped; midrib, strong; a pair of subprimaries branch from the midrib, 2 cm. above the base of the decurrent lamina, at an angle of 45°, and pursue a straight course to the apex of the lobes; 5 to 6 secondaries branch abaxially from the outer primaries at nearly 90° with the midrib and enter the marginal teeth; 5 to 6 pairs of opposite secondaries branch from the midrib at 50°, slightly curved upward and craspedodrome to the large marginal teeth; in the upper part of the leaf there are two sizes of marginal dentations, small dentations tend to alternate with the larger; 2–3 small secondaries branch from the inner side of the outer primaries and curve upward into the marginal teeth; tertiary veins weak, essentially reticulate but with a few percurrent cross-ties; fourth and fifth order veins merge to form a fine, quadrangular mesh, the spaces from .03 to .05 cm. in width; texture firm to coriaceous.

In spite of Brown's admonition quoted, it is necessary to name these fossils as a new species. The characters are, in many respects, intermediate between those of *Platanus raynoldsii* Newberry (see Brown, 1962:64, pls. 30–31) of the Paleocene and the living *P. lindeniana* Mart. and Gal. from eastern Mexico. The fossil closely resembles *P. raynoldsii* in gross features but differs in the wider spread of the leaf, the more marked lobes, and the finer areolation. The width of the fourth and fifth order meshes in the Paleocene species is from .07 to .09 cm., nearly twice that of *P. intermedia*. It is also similar to *P. appendiculata* Lesquereux from the early Eocene auriferous gravels on the west slope of the Sierra Nevada. The Wyoming species may be distinguished by the greater number of secondaries above the pair of side-primaries and by the weak, reticulate tertiaries. In *P. appendiculata* there are strong, percurrent tertiaries.

Occurrence.—Kisinger Lakes, Tipperary, Ramshorn. Type PA 5670.

Family SALICACEAE
Genus *Populus* Linné
Populus quintavena, new species
(Pl. 5, fig. 3)

Description.—Leaves round to round-ovate; length 9–10 cm., width, 7–8 cm.; base rounded to truncate; apex acute; petiole slender, flattened, complete length not preserved; margin evenly and finely crenate, approximately 4 rounded, projecting crenations per cm.; midrib straight, slender; two pairs of acrodrome veins arising from the top of the petiole, separated by angles of 25° to 35°, one to three pairs of secondaries in the upper third of the midrib, arising at 80°, strongly curved upward, approaching parallelism with the midrib; the inner pair of acrodrome veins gently bowed marginally, extending into the upper third of the leaf, bearing in their upper two-thirds 5 strong tertiaries arising at 45° and spaced approximately 1 cm., curved upward along the margin; the outer, acrodrome pair of veins extending not more than one-third the length of the lamina, spaced 1 cm. or less within the margin and bearing 6 to 9 tertiaries on the marginal side; the extremities of all these veins strongly looped along the margin and anastomosing by tertiary branches; veinlets from the apices of the marginal loops, 2 mm. or less within the margin, enter the glandular marginal crenations, others enter the intervening sinuses; irregular, flexuous cross-ties spaced 2 to 3 mm. apart bridge the spaces between the midrib and the side veins; major part of the tertiary venation consisting of a coarse, angular mesh; within this mesh is a finer, mostly pentagonal network, the opening 0.5 mm. in dimension; no further details preserved; texture, firm; frequency, rare.

Discussion.—These fossil leaves exhibit a certain resemblance to the foliage of the living *Cercidiphyllum japonicum* L. but differ in many respects such as the evenly crenate margin, the flexuous tertiary cross-ties, and the branching and fine marginal looping of tertiary veins. Although poplar leaves with five primaries are not common, it is with several species of oriental poplar that definite correspondences are found—*P. lasiocarpa* Oliver native to southwestern China and *P. ciliata* Wallich of northern India for example. The fossil leaves differ from those of the existing species in the less prominent glands of the dentations, in the fewer secondaries arising from the midrib, and in the greater symmetry of the dentations. The finer details of venation are remarkably similar. The living *P. lasiocarpa* is found in the mixed mesophytic forest of eastern Szechuan in company with *Fagus, Euptelea, Davidia, Acer, Pterocarya* and the like about 2400 m. down to 1200 m., with several species of oaks at the lower elevations (Wang, 1961:112–113).

The taxonomy of the fossil species of *Populus* and *Cercidiphyllum* is still in considerable confusion. Brown (1939) lumped together species of *Cocculus, Cissus, Populus, Cercidiphyllum* and probably other genera in his discussion of the fossil species of *Cercidiphyllum*. Wolfe has solved some of these taxonomic problems (1966:9–12), but some remain unsolved. Among fossil species *P. quintavena* is similar to *P. amblyrhncha* Ward and *P. flexuosa* Hollick (1936:63) from the Alaskan Tertiary and *P. xantholithensis* Knowlton (1899: pl. 85) from Yellowstone Park. The assignment of the first two species is somewhat doubtful, but the material from Yellowstone Park appears correctly assigned to *Populus*.

Occurrence.—Kisinger Lakes. Type PA 5613.

Populus wyomingiana (Berry), new combination

(Pl. 3, figs. 2, 3; pl. 7, fig. 3; pl. 9, figs. 2, 4; pl. 20, figs. 2, 3; pl. 21, fig. 3)

Zizyphus wyomingianus Berry, U.S. Geol. Surv. Prof. Paper 165, p. 73, pl. 11, figs. 6, 7, 1930.
Grewiopsis wyomingensis Berry, *ibid.*, p. 73, pl. 8, fig. 6, pl. 13, figs. 1, 2.

Description.—Leaves ovate; length 5 to 12 cm., width 3 to 6 cm., 1/w ratio 2.6; apex acute; base cuneate or, rarely, rounded, often weakly inequilateral; petiole slender, length 4 cm.; margin denate with rounded, glandular, curved teeth, prominently developed in the upper half of the leaf, the dentations tending to be somewhat irregular, occasionally of two sizes; midrib slender; a pair of acrodrome secondaries arise just above the top of the petiole at angles of 25°, ascending for a distance of approximately ⅔ the length of the lamina, ultimately approaching the margin and giving off tertiary branches to the marginal teeth; marginal side of the acrodrome secondaries marked by tertiary branches approximately at right angles to the midrib, looping along the margin and branching to the marginal dentations; rarely a second pair of thin secondaries arise at the base of the acrodrome pair; 3 to 4 pairs of curving secondaries arising at wide angles, 50° to 60°, in the upper half of the leaf, branching and looping along the margin; midrib marked by transverse veins of tertiary order which connect with the acrodrome pair of secondaries; tertiary venation reticulate with coarse meshes roughly elongated perpendicular to the midrib; quartary venation reticulate, meshes averaging 1 mm. in size; quintary meshes 0.2 to 0.3 mm. in dimension, within which are simply branched free endings; texture firm; frequency, abundant.

Discussion.—Notwithstanding their superficial resemblance to the foliage of several genera showing acrodrome venation such as *Cercidiphyllum, Colubrina, Banara, Grewia* and *Zizyphus*, the curving, glandular dentations, the craspedodrome tertiary branches, and ultimate, free-ending veinlets, show that these fossil leaves cannot be assigned to any of these genera. Some of the leaf characters also suggest the foliage of *Aphananthe, Celtis* and *Luehea*. The marginal characters of *Aphananthe* and its craspedodrome secondaries have no correspondences with the fossil leaves. The fine venation of *Celtis* is entirely different, since the ultimate veins have club-shaped endings or end in round dots or thickenings. The leaves of *Luehea* are characterized by a fine, even-textured, closed mesh, not found in these fossils. The characters of the margin and all details of venation are matched by the foliage of *Populus;* the correspondences with the areolation are exact. The marked acrodrome venation is not commonly found in the leaves of *Populus* but a close examination shows that these veins are truly secondaries and not sub-primaries. The living *P. ciliata* Wallich from northern India has foliage which shows marked correspondences with the fossils. Leaves of *P. heterophylla* L. of the eastern states possess the typical margin and major details of venation, although the apex in that species is often blunt. The finer details of nervation are found in the leaves of *P. dimorpha* Brandegee of northwest Mexico. This species appears to be a comparatively ancient or early member of the genus. A few leaves of *P. wyomingiana* possess an extended apex similar to, but not as pronounced as that of the Green River species *P. wilmattae* Cockerell. Since *P. dimorpha* is the living species most similar to the Green River *P. wilmattae* this relationship suggests that *P. wyomingiana* was ancestral to the Green River and existing northwest Mexican species. *P. cinnamomoides* of the Green River flora also resembles the small, narrow forms of *P. wyomingiana,* which indicates that this Green River species was either derived from *P. wyomingiana* or that the two represent branches from a common ancestral type. This ancestral species is most probably a new species in the Wind River flora of Lost Cabinian age now under study by Dorf, Hickey and MacGinitie. The Wind River species is similar to that from the Kisinger Lakes locality but differs in the less prominent marginal dentations and in the less complex branching of the free ending veinlets. It is possible that these two forms represent the same species and that the differences are only ecologic in rank.

A species of *Populus* from the Mormon Creek flora, called *Cercidiphyllum elongatum* Brown by Becker (1960: pl. 29, figs. 1–6) seems clearly a derivative of *P. wyomingiana,* if not the same species. The age of the Mormon Creek flora has been in dispute. Becker assigned it to the late Lower Eocene but studies in progress of the Lost Cabinian floras of the western Wind River Basin indicate that this cannot be correct. The generic and specific composition of the Wind River floras show that the Mormon Creek flora must be younger than early Eocene. It is also younger than the Kisinger Lakes-Tipperary flora since the species of *Amelanchier, Crataegus, Koelreuteria, Mahonia, Acer* and *Ptelea* place it closer to the Florissant, although it is clearly older than the Florissant. I am therefore led to the conclusion that the most probable age of the Mormon Creek flora is Late Eocene. If this is correct then this flora is the only Late Eocene megaflora now known from the entire area of the middle Rockies.

We have then a sequence of poplar species extending through the Paleogene of the middle Rockies: (1) new species from the Wind River flora, late Early Eocene, (2) *P. wyomingiana* and *P. quintavena* from the Kisinger Lakes-Tipperary flora, basal Middle Eocene, (3) *P. wilmattae,* and *P. cinnamomoides* from the Green River flora, Middle Eocene, and (4) new species from the Mormon Creek flora, Late Eocene?

P. wyomingiana differs from *P. quintavena* in several respects; in *P. wyomingiana* are found (1) the strong, acrodrome side-veins arising at small angles, (2) the numerous tertiary cross-ties between the lower midrib and the acrodrome secondaries, at angles nearly perpendicular to the midrib, (3) the tertiary meshes elongated across the leaf nearly perpendicular to the midrib. The last two characters appear to be relatively primitive in the leaves of *Populus.*

Berry (1930a:106, pls. 15, 16) described and figured fossil leaves which he named *Grewiopsis tennesseensis* from what are probably early Middle Eocene deposits of the Gulf Coast. These appear to be fossils of *Populus* leaves similar to *P. wilmattae.* This similarity was overlooked in the discussion of *Populus* in the Green River flora (MacGinitie, 1969:93).

There are approximately two dozen existing species of poplar confined to the north temperate regions of the world. The earlier Eocene in the middle Rockies seems to have been the beginning of active speciation in the genus, and this speciation continued through the Tertiary. *Salix,* a closely related genus did not evolve as rapidly until the Neogene and speciation in the willows seems to be still proceeding at a relatively rapid rate. In the Paleogene the genus *Populus* produced several species whose existence covered a relatively short time span and these are useful stratigraphic markers. The more advanced type of nervation appears to be that typified by *P. heterophylla,* the relatively coarse, irregular, tertiary and finer network; while the relatively formalized type of *P. dimorpha* with the tertiary venation tending to be aligned perpendicular to the midrib is probably the more primitive.

P. balsamifera L. and *P. tremuloides* Michaux extend northward of the 65th parallel in northwestern Canada to the border of the tundra, while *P. dimorpha* Brandegee occurs as far south as the 27th parallel on the west coast of Mexico. Some species, such as *P. trichocarpa* Torrey and Gray show a remarkable range of temperature adaptation. This species is found from the Kenai peninsula of Alaska to the mountains of Baja California, from latitude 62°N to 31°N. Adaptations of the genus throughout the Tertiary seem to have been toward colder climates. Although some living species, *P. grandidentata* Mich. for example, are typical of upland situations, plants of the genus, in general, favor moist soils and are characteristic of floodplain and lakeside habitats.

Occurrence.—Kisinger Lakes, Tipperary. Hypotypes PA 5607, PA 5622, PA 5628, PA 5629, PA 5682, PA 5684.

Family PROTEACEAE ?

Genus **Proteaciphyllum,** new genus
Proteaciphyllum minutum, new species
(Pl. 5, fig. 2; pl. 18, fig. 2)

Description.—Leaves (or leaflets) ovate-lanceolate, length 1.7 to 2.2 cm.; width 5 mm.; base cuneate to acute; apex acute, sessile or nearly so, a heavy, cross-ridged 1×1 mm. attachment at the base, margin entire; venation parallelodromous, midrib slender flanked by two delicate, sinuous primaries at angles of approximately 10°, the primaries extending to near the apex, approaching parallelism with the margin and converging at the apex; primaries connected by numerous, elongated, irregular, thin secondaries, nearly parallel to the margin, forming elongated meshes 0.3 to 0.9 mm. in length, secondaries at higher angles on the marginal side of the primaries, up to 45°, forming pronounced loops; a series of fine, delicate loops elongated just within the margin, the outer sides of the loops forming an irregular marginal vein; finer venation obscure; texture firm; frequency, rare.

Discussion.—These strange, small leaflets or leaves recall the foliage of certain species of *Mahonia, Acacia,* and *Guaiacum* but the peculiar marginal looping excludes them from any real relationships to these genera. The details of venation, shape of the leaves, and the near sessile base, correspond with those characters of several species of *Hakea,* particularly *Hakea dactyloides* Cavanilles, and *Hakea crassifolia* Meisner of southern and southwestern Australia. A genuine relationship to *Hakea* or others of the Proteaceae may be questioned in view of the modern distribution of the genus. However the characters of venation are remarkably similar. It is possible that these fossils represent detached sepals or petals of some unknown genus. The genus *Proteaciphyllum* was chosen to indicate likeness to foliage of the Proteaceae without implying undoubted relationship.

There have been nearly 100 species of *Hakea* described from Australia. The plants are shrubs or trees with xeric adaptations.

Occurrence.—Kisinger Lakes. Type PA 5612; paratype PA 5673, topotype PA 5614.

Family ROSACEAE

Genus *Prunus* Linné
Prunus nota, new species
(Pl. 2, fig. 3; pl. 3, fig. 1)

Description.—Leaf broadly lanceolate, length 8 cm., width 2.2 cm.; apex extended to an acuminate tip; base narrowly rounded, greatest width in the upper one-third of the lamina, a pair of petiolar glands 2 mm. below the lamina; petiole stout, 1.7 cm. in length; margin finely and regularly crenate-dentate, the dentation prominently gland-tipped, 10 dentations per cm.; midrib slender; ten pairs of alternation secondaries, originating at 45°, slightly curved upward (obarcuate) and forming prominent loops along and 2–3 mm. within the margin, the first order loops succeeded by at least 4 other orders; tertiaries finely and irregularly percurrent, the veinlets aligned perpendicular to the midrib, and spaced 1 mm. or less; quartary veins nearly perpendicular to the tertiaries forming a complex, mostly quadrangular mesh, the openings 0.5 to 0.8 mm. in dimension, slightly elongated perpendicular to the midrib; finer details of nervation not visible; texture firm; frequency, rare.

Discussion.—The fossil leaf is much like the leaves of the living *Prunus serotina* Ehrlich, the black cherry of the forests east of the Mississippi. The characteristic margin, the streaming-reticulate nervation oriented perpendicular to the midrib, and the pair of acropetiolar glands prove the assignment of this fossil to the genus *Prunus.* The existing *P. serotina* is a large tree of the "summer-green" (deciduous) forest of eastern North America, where it characteristically occurs scattered in the forest of beech, sugar maple, tulip tree, ash, and white oak, although it is found throughout the eastern forests from southern Canada to the Gulf border. A variety, sometimes called *P. capuli* Cavanilles, is common in Mexico, especially in the eastern Sierra Madre.

Only one complete leaf of this species has been found, suggesting that the tree grew at some distance from the site of deposition, perhaps at higher elevations toward the upper reaches of the stream. A fossil species similar to *P. nota* has been described by me from the Rocky Mountain fossil floras; *P. stewartii* (*gracilis*) from Florissant (MacGinitie, 1953:120) and the Green River flora (MacGinitie, 1969:105) in northwestern Colorado. The two species appear to be related, but the fossils are so rare it is not possible with the material at hand to establish any species identity.

Occurrence.—Kisinger Lakes. Type 5606.

Family PALMAE
Genus *Sabalites* Saporta
Sabalites florissanti (Lesquereux) Berry

Sabalites florissanti (Lesquereux) Berry, U.S. Geol. Surv. Prof. Paper 165, p. 66, pl. 9, 1930.
Falbellaria florissanti Lesquereux, U.S. Geol. Surv. Rept. Terr., vol. 8 (The Cret. and Tert. Floras), p. 144, pl. 24, figs. 1–2a, 1883.
Sabalites powelli (Newberry) Berry, U.S. Geol. Surv. Prof. Paper 165, p. 67, pl. 10, figs. 6, 7, 1930.
Sabal powellii Newberry, U.S. Geol. Surv. Mon. 35, p. 30, pl. 63, fig. 6; pl. 64, figs. 1–1a, 1898.

Genus *Geonomites* (Visiani) Lesquereux
Geonomites haydenii (Newberry) Berry

Geonomites haydenii (Newberry) Berry, U.S. Geol. Surv. Prof. Paper 165, p. 67, 1930.

Genus *Manicaria* Gaertner
Manicaria haydenii Newberry

Manicaria haydenii Newberry, U.S. Geol. Surv. Mon. 35, p. 31, pl. 64, fig. 3, 1898.

Genus *Palmocarpon* Lesquereux
Palmocarpon lesquereuxi Berry

Palmocarpon lesquereuxi Berry, U.S. Geol. Surv. Prof. Paper 165, p. 67, 1930.

In many instances the identification of palm genera based on fossil material is not feasible since palm leaves do not show good generic or specific characters, excepting cuticular morphology. Reasonably good taxonomic assignments may sometimes be made if the cuticle is preserved. Thus, in the absence of well-preserved cuticle, it is best to use form genera: *Sabalites* for the common fan palms, and *Geonomites* for the feather-type palms. Therefore Berry's original names have been kept.

It is possible that several species of fan palms were present in the area when these fossils were deposited. Berry distinguished two types on the basis of an extended acumen as contrasted to a blunt or rounded acumen. On the basis of the characters preserved, the fossil leaves are much like the foliage of the existing *Sabal palmetto* Lodd. Some of the fossil palm leaves show a thorny or armed rachis (petiole) similar to that of the living *Washingtonia filifera* Wendlich, the fan palm of the California desert. There are about ten living species of Sabal ranging through warm temperate to tropical America, of which two or three are found in the United States from Florida to the North Carolina coastal islands. *S. palmetto* is characteristic of low ground "prairies," marshes, pine lands and hammocks of the coastal plains.

At the Boysen locality north of Shoshoni, Wyoming, there are the impressions of thousands of palm leaves with the minute details of venation preserved. The leaves were large, up to three feet in diameter or larger. The venation consists of a strong midrib for each palmate division, flanked by an average of 16 longitudinal veins, spaced 12 per cm. Between the main longitudinal veins are one to four much finer longitudinal veins. The strong veins are connected by irregular cross nervilles spanning one or two spaces and spaced approximately 3 mm. longitudinally. This type of venation is found in the palm genera *Brahea, Serenoa* and *Sabal*. In *Serenoa* the cross-nervilles are nearly continuous across the half-blade and thus the fossils are not that genus. The inter-veins are more numerous and finer in *Brahea* although the leaves of that genus are close to the fossil impressions. The characters of these fossils are closer to those of *Sabal* leaves than to those of any other genus examined. Thus the Kisinger-Tipperary fan palms probably represent species of *Sabal*. The associated species show that these palms grew on low, moist ground.

There are only fragments of the feather palm preserved and thus no clear relationship to any living species can be determined. The living genus *Geonoma* to which the fossils have been likened comprises 80 to 90 species which are confined to tropical America and the West Indies. In Mexico species of *Geonoma* are found in the Tierra Templada at considerable elevations in the pine-oak forests and the typical mixed deciduous forests.

Palm leaf impressions are abundant in many of the Middle Eocene floras of the central Rockies, but relatively uncommon in the Early Eocene strata. A striking example of the vegetational changes through time is illustrated while collecting the rich, semi-tropical Middle Eocene floras with abundant palms in the central and eastern Wind River Basin with its present impoverished, cold, semi-desert, steppe flora, severe winters and short growing season. This is a contrast which can best be appreciated by collecting. The worker excavating the Eocene floras is carried back into another world in which conditions had little resemblance to those of the present.

To conclude, the Kisinger-Tipperary flora contains at least two species of fan palm similar to the *Sabal* palms of the southeast coast, and a feather palm similar to *Geonoma* mexicana Liebmann of the Mexican Tierra Templada.

Remains of fan palms are relatively common in the Tertiary of the Gulf Coast. These have been called *Sabalites grayanus* Lesquereux (see La Motte, 1952:311) or *Sabalites vickburgensis* Berry (1916:233). It is probable that several species are also involved here. Fossils of *Sabalites* are common in the floras of the central Rockies from the late Cretaceous through the middle Eocene. It is not possible from the published work to determine whether any of these are conspecific with the *Sabalites* from the Wind River Basin Eocene. *Sabalites californicus* Lesquereux from the Early Eocene of Chalk Bluffs and La Porte, California (MacGinitie, 1941:99) differs from the Wind River species in the wider spacing of the secondaries, and in the more numerous and longer cross-tertiaries. This species is more like the foliage of the living *Rhapis* of southeastern Asia than that of *Sabal*.

Occurrence.—Kisinger Lakes, Tipperary, Boysen. Hypotype PA 5721.

Family SALICACEAE

Genus *Salix* Linné

Salix molesta, new species

(Pl. 4, figs. 1, 2, 3)

Description.—Leaves linear to linear-lanceolate; length 7 to 18 cm.; width 2.4 to 3.5 cm.; apex long-acuminate; base narrowly cuneate; petiole relatively stout, maximum length 2 cm.; midrib strong; margin with close-set, rounded, glandular dentations, 6 per cm. in the larger leaves; secondaries numerous, opposite to subopposite, spaced 2.5 per cm., originating at 35° to 45° in the central portion of the lamina, curved upward at about half the distance to the margin, ascending near the margin forming a succession of fine, complex loops; with numerous cross-ties to the succeeding secondary tertiary venation coarsely reticulate, but with scattered percurrent cross-ties; quartary venation irregularly reticulate, the meshes 0.4 to 0.5 mm. in dimension; areolation a fine, uniform mesh whose openings are from 0.1 to 0.15 mm. in greatest dimension, free endings not observable; texture firm; frequency, common.

Discussion.—The long petioles, finely rounded-denate margin and the ascending secondaries give these fossil leaves the appearance of large *Salix cockerellii* leaves (Brown, 1934:53, pl. 9), but the secondaries in that fossil species are more numerous, sub-parallel and not so ascending, and with a greater angle between the secondaries and midrib. The material from Tipperary and

Kisinger Lakes is clearly a different species, although the Green River form may have been derived from this species. The combination of characters given above distinguish this willow leaf from any previously described from North America.

Among living species the fossil most resembles the foliage of *Salix bonplandiana* H. B. K. from Mexico and *S. chilensis* Molina from Peru. The fossil differs from all the modern species examined in the numerous rounded dentations, combined with the elongate form and strongly ascending secondaries. There is also some resemblance to the foliage of the existing *Salix nigra* Marshall but the fossil does not possess the marginal "hem" formed by the anastomosing secondaries as in the living species.

Occurrence.—Kisinger Lakes, Tipperary. Type PA 5609; paratype PA 5608.

Family SALVINIACEAE
Genus *Salvinia* Adanson
Salvinia preauriculata Berry

Salvinia preauriculata Berry, U.S. Geol. Surv. Prof. Paper 165, p. 62, pl. 6, figs. 1–3, 1930.

Although Berry states that "remains of a species of *Salvinia* are abundant in these deposits" (Tipperary), and the figures seem to be clearly leaf fossils of this genus, none have been found during recent extensive collecting. It is possible that Berry confused the foliage of *Spirodella* with the leaves of *Salvinia*. Thus, for the present, the occurrence of *Salvinia* at Tipperary is probable but has not been proved. It may be that Berry's fossils of *Salvinia* came from Lenore or Crowheart Butte, Wyoming, since he studied the three collections at the same time. The original material is in the paleobotanical collection in the National Museum; the matrix seems to be the same as that at Tipperary and thus the species is accepted as occurring at that locality.

Salvinia with *Azolla* comprise this family of highly specialized, floating, water ferns, *Salvinia natans* (L.) Allioni is found living in southern France, the southern United States, India and China, and thus is relatively hardy to freezing, but the genus as a whole is largely confined to the tropics.

Occurrence.—Tipperary.

Family DILLENIACEAE
Genus *Saurauia* Willdenow
Saurauia propia, new species
(Pl. 31, figs. 1, 2, 3)

Description.—Leaf broadly lanceolate; apex acuminate; base cuneate; length 15 cm., width 4 cm.; petiole heavy, length unknown; margin finely and irregularly serrate, approximately 5 dentations per cm., the serrations falcate in shape, often glandular, and rounded rather than sharp; midrib heavy; secondary veins numerous, 25 or more, opposite, arising at 60°, nearly straight for ¾ their length, then abruptly curved upward and approaching the margin, giving off tertiary branches to the serrations and to the next succeeding secondary; each pair of secondaries separated by an inter-secondary extending about ⅔ of the distance to the margin and terminating in a series of tertiary branches; marked parallelism of secondary nerves; tertiary venation of numerous, fine, percurrent veinlets, aligned across the leaf, the angle with the midrib from 10° to 20° below the horizontal, veinlets often showing bifurcating branches toward the margin, each marginal dentation receives a tertiary branch from the outer side of the secondary, tertiaries spaced 11–12 per cm. in the middle of the lamina; quartary venation of slender veinlets, branching at high angles from the tertiaries and forming a reticulate mesh with openings of irregular sizes; texture coriaceous; frequency, rare.

Discussion.—The lanceolate shape, serrate margin, sub-parallel secondaries and inter-secondaries, and the peculiar tertiary nervilles aligned across the leaf are characteristic of the leaves of *Saurauia*. The genus comprises about 250 species native to the tropics of America and Asia and typical of the tropical rain forests (Wang, 1961:161–163). The fossil leaf shares the characteristics of the leaves of several living species of *Saurauia* such as *S. belizensis* Lundell of Central America, *S. singalensis* Karth of southern Asia and *S. elegans* (Choisy) F. Viel of the

Philippines. The fossils differ from all the living species examined in the greater number and parallelism of the secondaries. They appear to share a greater number of characters with the tropical American species.

There are certain resemblances to the leaves of *Curatella* and *Ouratea,* but the differences prohibit an assignment to either of those genera. Since large fossil leaves of *Salix* are found in the flora, the question may arise why do not these fossils represent the leaves of *Salix,* due to rather marked but superficial resemblances? The following characters, among others, serve to distinguish the leaves of the two genera. In *Saurauia* the secondaries are slightly decurrent at the juncture with the midrib; in *Salix* the secondaries meet the midrib directly at high angles, 70° or more. In *Salix* the secondaries curve upward toward the margin for nearly half their length where they connect by tertiaries with the succeeding vein; in *Saurauia* the secondaries approach much closer to the margin before looping and sending tertiary branches to the serrations. In *Saurauia* the tertiaries are nearly perpendicular to the midrib and form a marked pattern of fine veins streaming across the leaf. In *Salix* the tertiaries tend to be nearly perpendicular to the secondaries and they form a reticulate net not oriented across the leaf. The fossil leaves of *Saurauia* show scattered dots, tiny thickenings, along the veins. The *Salix* leaves used for this comparison are those of *S. bonplandiana* H. B. K. since these are most like the fossil leaves of *Salix* found in the flora.

Oreopanax mississippiensis Berry (1924:88, pl. 22, fig. 3) from the Gosport sand of the Gulf Coast has characters of both *Saurauia* and *Ouratea.* The resemblance to the Kisinger Lakes fossils indicate a relationship, but this cannot be verified on the basis of the fossil material available.

Occurrence.—Kisinger Lakes. Type PA 5723; Paratype PA 5722.

Family ARALIACEAE

Genus *Schefflera* Forster

Schefflera insolita, new species

(Pl. 28, figs. 1, 2)

Description.—Leaflet lanceolate; length 13 cm., width 2 cm., greatest width in the lower third; petiolule not preserved; apex attenuate; base cuneate; margin entire; midrib strong; 14 pairs of irregularly disposed secondaries, arising at 45° in the central portion of the lamina, gently curved upward, ascending and approaching parallelism with the margin, at 1 mm. within the margin numerous small loops formed with tertiary branches from the succeeding secondary, disappearing in the tertiary network but not merging with the margin; tertiary venation streaming-reticulate the veins oriented nearly perpendicular to the midrib across the lamina, tertiary veins spaced approximately 1 mm. along the midrib; quartary meshes elongated perpendicular to the midrib, the individual areoles occupied by rather complexly branched, free-ending veinlets; texture heavy; frequency, rare.

Discussion.—The ascending secondaries and the tertiary meshes oriented perpendicular to the midrib indicates that this fossil leaflet represents a genus of the Araliaceae. The most marked correspondences are with the leaflets of *Schefflera,* particularly *S. heterophylla* Clemens of tropical southeastern Asia and the East Indies.

No closely similar fossil leaflet has been described from the American Tertiary. The species of *Schefflera* are small trees confined to lowland habitats in the tropics of both hemispheres. The large leaves are digitately compound with separate, usually long-petiolulate leaflets. At present the genus is adapated to tropical environments only.

Occurrence.—Kisinger Lakes. Type PA 5713; paratype PA 5712.

Family SAPINDACEAE

Genus *Serjania* Plumier ex Schumacher

Serjania rara, new species

(Pl. 25, figs. 1, 2, 3)

Description.—Leaflets of two forms, terminal and lateral. Terminal leaflet obovate, palmately lobed; length 6 cm., width 4 cm.; apex acute; base cuneate-decurrent, sessile; margin with a few, paired coarse dentations, the lateral lobes projecting 7 mm. from the margin, defined above by wide, sharp-angled sinuses; midrib slender; 4 pairs of subopposite secondaries, originating at from 40° to 50°, the lower pair stronger and entering the marginal lobes, alternately craspedodrome or proceeding to the sinus where they bifurcate and loop along the margin, both above and below the sinus; tertiary venation of many, irregular, curving cross-ties, bowed abaxially, aligned nearly perpendicular to the midrib, 3 to 5 tertiaries between each secondary, originating along the midrib, 4 or 5 strong tertiaries on the marginal side of the lobar secondaries, curving upward, looping and finally merging into the margin; quartary venation an irregular net, the meshes averaging 1 mm. in greatest dimension and bounded by sinuous, angular veinlets; quintary venation obscure but apparently a fine, irregular net, the meshes 0.2 to 0.3 mm. in dimension; texture firm. Lateral leaflets markedly asymmetric; midrib slender; length 2.5 to 3.5 cm.; width 1.5 to 2 cm.; apex rounded to acute; base broadly cuneate; margin with 3 to 4 large dentations, more marked on one margin (outer); 5 to 6 pairs of secondaries, alternately craspedodrome and camptodrome; remainder of venation as in the terminal leaflet; texture firm; frequency, not rare.

Discussion.—These fossils display the characteristic features of the foliage borne by several living species of *Serjania*. The cuneate base of the terminal leaflet, the strong, relatively blunt dentations, the alternate craspedodrome and camptodrome secondaries, and the peculiar quartary mesh are diagnostic of the foliage of the genus. The fossil leaves are similar to the foliage of several species of *Serjania* now growing in tropical North America; *S. rhombea* Radlkofer of Panama, *S. racemosa* Schumacher of Mexico and Central America, and *S. triquetra* Radlkofer of Costa Rica. The genus is said to comprise about 175 species, and is confined to tropical America. The plants are strong growing lianas with compound leaves and 3-winged schizocarps. They are common in the deciduous and semi-deciduous forests of the west coast of Mexico south of Mazatlan.

Occurrence.—Kisinger Lakes. Type PA 5697; paratypes PA 5696, PA 5698; topotypes PA 5700, PA 5701, PA 5702.

Family SPARGANIACEAE
Genus *Sparganium* Linné
Sparganium antiquum (Newberry) Berry
(Pl. 7, fig. 1; pl. 14, fig. 2)

Sparganium antiquum (Newberry) Berry, U.S. Geol. Surv. Prof. Paper 165, pp. 64–65, pl. 8, fig. 5, 1930.

Berry discussed this species at length and gave a restoration in terms of the material available to him. The assignment to *Sparganium* appears to be correct. However, Berry's restoration (1930:65) is in error. These fruiting heads were borne in an elongated, loose raceme, whose length was from 15 to 20 cm., on peduncles from 1.5 to 2.5 cm. in length, similar to the branching inflorescence of the living *S. eurycarpum* Engelmann which is widespread in North America. The characters of these fossils are matched in every respect by the fruiting heads of the bur reed (*Sparganium*); the numerous, small beaked fruits, radially arranged and interspersed by chaffy perianth scales are clearly preserved. Since these fossils are always associated with leaf impressions of *Platanus* or *Platanophyllum* this has led some authorities to suggest that the fruiting heads might be those of *Platanus*, but they are clearly unlike the fruits of any living species of that genus.

In the Eocene of the central Rockies fossils of this *Sparganium* are invariably accompanied by leaf fossils of *Platanus* and *Lygodium*. In fact a flora containing abundant *Lygodium*, *Platanus* cf. *brownii* and *Sparganium* heads is almost sure to be of Middle Eocene age. The species is common in the Wind River flora of Lost Cabinian age and also in the Green River flora.

Sparganium is the only genus of the family. There are from 12 to 18 species confined to the north temperate regions, Australia and New Zealand, a distribution suggesting considerable antiquity. The species are confined to aquatic or semi-aquatic habitats.

Occurrence.—Tipperary, Kisinger Lakes. Hypotypes PA 5624, PA 5652.

Family SAPINDACEAE
Genus *Sapindus* Linné
Sapindus dentoni Lesquereux

Sapindus dentoni Lesquereux, U.S. Geol. Surv. Terr., vol. 7, p. 265, pl. 64, figs. 2–4, 1878; Berry, U.S. Geol. Surv. Prof. Paper 165, p. 71, 1930.

The original specimens of *"Sapindus dentoni"* cannot represent the genus *Sapindus*. They were almost certainly leaves of *Zelkova* with poorly preserved or undulate margins. In spite of this the species has been cited from many localities of the western Tertiary by Knowlton, Berry, Ball, Brown and others. Each occurence will need to be reassigned to its proper taxonomic position. Some of the occurrences may be fossils of *Sapindus*, others are, at present, of unknown affinities. The original types are no longer to be found. Thus *Zelkova nervosa* (Newberry) Brown from the Green River Formation of northwestern Colorado, should probably be renamed *Zelkova dentoni*, but, in view of the missing types the best procedure appears to be an avoidance of this change.

Family LEMNACEAE
Genus *Spirodela* Schleiden
Spirodela magna, new species
(Pl. 6, fig. 4; pl. 21, fig. 4)

Description.—"Leaves" orbicular, 1 to 1.5 cm. in diameter, seeming often to be paired, the pair consisting of a larger, orbicular lamina subtended at the base by a smaller (reproductive bud?), round or oblong form, 1 cm. in length by 5 mm. in width; venation of numerous (10 to 14) unbranched veins, arising from the attachment point of the roots, ascending to the margin, the central vein nearly straight, but the veins progressively more curved toward the edge of the lamina until the outer veins are curved approximately parallel with the margin, the extremities of the veins strongly curved toward the center of the lamina, finally merging with the margin; between the veins is a complex, but even-textured network of veinlets, the areoles approximately 0.2 mm. in dimension; margin entire; texture firm, frequency, common.

Discussion.—These small impressions correspond in all characters with those of the living *Spirodela polyrrhiza* (L.) Schleiden with the exception of size. The fossils average twice the size of any known species of the genus. This one difference seems insufficient to warrant creation of a new genus, since all other characters correspond. Brown (1962: pl. 16) figures *Hydromystria expansa* (Heer) Hantke from the Paleocene of the Rocky Mountains in Montana, Wyoming and North Dakota. These are similar to the Kisinger fossils but show differences in shape and venation. The Fort Union fossils may belong to the same family but are probably not the same genus. Brown questioned the assignment of *Hydromystria* and discarded the resemblance to *Lemna* on the basis of size.

Spirodela scutata (Dawson) Berry, from the Paleocene of British Columbia, Alberta and Montana (Dawson, 1875:329, pl. 16, figs. 5, 6) has a remarkable resemblance to the Wyoming fossils. This species was first called *Lemna* by Dawson but was changed to *Spirodela* by Berry (1935:23). Bell (1949:82, pl. 63, figs. 1, 3) has illustrated some of the fossil material from Alberta, and they correspond closely with the characters of the *Spirodela* from Kisinger Lakes. The original types are not available to me and so the decision as to specific identity cannot be made. A time interval of approximately 15 million years separates the two species, and, although it is possible that they might be identical, it appears best to keep them as separate species until comparisons can be made between adequate fossil suites.

The genus is a much-reduced and specialized plant which floats on the surface of still ponds, sometimes practically covering the surface. It has no true leaves or branches, but resembles algae in its simplified plant body. There are no more than two species which are almost cosmopolitan (none in Africa ?). *S. polyrrhiza* is found everywhere in the Americas from Canada to Patagonia. The genus thus has no particular climatic significance.

Occurrence.—Kisinger Lakes. Type PA 5617; paratype 5686; topotypes PA 5618, PA 5621.

Family STERCULIACEAE
Genus *Sterculia* Linné
Sterculia subtilis, new species
(Pl. 30, figs. 1, 3)

Description.—Leaves ovate; length 11 to 15 cm., width 4 to 6.2 cm., greatest width in the middle of the lamina; apex acute; base rounded to incipiently cordate; petiole slender, 3.5 cm. preserved; midrib, slender, straight; 9–12 pairs of secondaries, opposite or subopposite, the basal pair pseudo-acrodrome arising at 45°, the succeeding pair arising at 70°, the angle decreasing toward the apex; margin entire or undulate, the undulate specimens with widely scattered, shallow dentations; secondaries essentially straight until approaching within 5 mm. or less of the margin where they curve sharply upward, ascending just within the margin and forming fine, complex serial looping by tertiary cross-ties with the succeeding secondary; the margin characterized by "stitching" formed by the outer loops; the marginal dentations entered by tertiary branches; many small tertiary branches on the outer sides of the secondaries; tertiary venation with scattered cross-ties but essentially reticulate, forming a thin, coarse mesh, the quadrangular openings from 1 to 1.2 cm. in greatest dimension; within the tertiary mesh is a fine, quadrangular ultimate mesh, the openings from 0.2 to 0.3 mm. in greatest dimension; indistinct branching within these meshes which may be free-ending; texture firm; frequency, not common.

Discussion.—These fossils may be recognized by the entire to undulate-dentate margin, the long petiole, the pseudo-acrodrome basal secondaries, and especially by the characteristic, fine-quadrangular areolation. This areolation is so distinctive, that small fragments of the fossil leaves can be easily recognized. The last orders of venation tend to merge one into the other. This venation is typical of the foliage of the family Sterculiaceae. The closest resemblance is with the venation of the genus *Sterculia* although the characters of *Commersonia* are also close. The fossils do not show the completely dentate margin of *Commersonia*. The characters of the fossil indicate that it is a comparatively ancient form whose foliar morphology was already established by the beginning of the Middle Eocene. The correspondences in venation between the fossils and the existing species are remarkable considering the time lapse. They differ from the living foliage in the undulate to widely dentate margin and in the less marked development of tertiary cross-ties; otherwise there appears to be no essential difference between the foliage of the existing and the Eocene species.

The fossils most resemble a group of species from tropical southeastern Asia, typified by *S. javanica* R. Blume, *S. lanceolata* Cavanilles and *S. oblongata* Brown. There are nearly a dozen species in this general area which are so similar as to suggest over-speciation. *S. pruriens* Schumacher from British Guiana also has the same type of foliage. The only species known to me whose leaves show the undulate-dentate margin is *S. pilosa* Ducke from the Amazon basin.

Potbury (1935:76) described a *Sterculia* leaf similar to that from Kisinger Lakes. Although it represents the same general type of foliage, Potbury's specimen differs in the fewer secondaries, stronger tertiaries, and greater development of the pseudoacrodrome, basal secondaries. The typical, fine areolation appears to be present. Berry described about a half dozen species of *Sterculia* from the Paleogene floras of the Gulf Coast, but these are all of the lobed type of foliage not closely similar to the Wyoming fossils.

The living species of *Sterculia* comprise small to large trees, typical of the tropical rain forests, although a few species in Asia extend into temperate regions. In the Amazon forests the plants often grow to heights of 120 feet or more. The foliage is variable; simple-entire, palmately lobed or palmately compound.

Occurrence.—Kisinger Lakes. Type PA 5719; topotypes PA 5720, PA 5720a.

Family SYMPLOCACEAE
Genus *Symplocos* Jacquin
Symplocos incondita, new species
(Pl. 9, fig. 1; pl. 17, fig. 1; pl. 20, fig. 1; pl. 32, figs. 1, 2)

Description.—Leaves ovate; length 4 to 11 cm., width 1.3 to 4.5 cm., length/width ratio 2.6 to 3.3; apex acute; base cuneate; margin entire or remotely and irregularly dentate in the upper third of the lamina, the upper margins of the small teeth are squared off nearly perpendicular to the midrib above (or apically) and are curved into the line of the margin basally, the teeth are entered by veinlets of tertiary or higher order; petiole stout, 8 to 11 mm. in length; venation dictyodromous; midrib strong; 6 to 10 subopposite secondaries arise at small angles, ascend nearly parallel to the midrib for 1 to 4 mm. curve strongly outward at angles of 45° to 50° with the midrib, bow upward and form complex loops 4 to 1 mm. within the margin, one or two inter-secondaries are spaced irrigularly between the secondaries; both sets of secondaries are thin and complexly branched, the first bifurcates about halfway between the midrib and the margin, continuing bifurcations and looping form a series of three successively finer loops near the margin, bounded by tertiary, quartary and quintary branches in order; tertiary venation a coarse, complex mesh, the meshes mostly quadrangular and averaging 1.5 mm. in width, these enclose a quartary mesh whose openings are 0.5 to 0.7 mm. in width. Within the quartary meshes are complexly branched (bifurcating 2–3 times), free-ending veinlets; texture coriaceous; frequency, abundant.

The peculiar, almost disorganized, venation of the fossil leaves, their cuneate bases and heavy petioles are characteristic of the foliage of several living species of *Symplocos*, particularly *S. nitens* Bentham of Paraguay, *S. crassifolia* Bentham and *S. setchensis* Brand of southern China, and *S. tinctoria* L'Heritier of the southeastern states. There is also some resemblance to the leaves of living species of *Osmanthus*, but the foliage of that genus lacks the complex looping of the secondary termini, the peculiar origination of the secondaries, and shows a larger angle of the secondaries with the midrib. The areolation is similar. The two genera are closely related, and foliar resemblances are to be expected.

Symplocos oregona Chaney and Sanborn (1933) from the Goshen flora of western Oregon appears to be the only fossil species of the genus hitherto described from the Tertiary of western America. This fossil foliage is somewhat similar to the material from Wyoming, although the margin is dentate throughout its length, and the angle of the secondaries is greater. The state of preservation of the leaves from Oregon does not permit comparison of the finer details of venation.

The genus *Symplocos* is widespread in both hemispheres, ranging from the tropics to warm temperate areas. There are approximately 290 living species. They are abundant in the moister areas of South America, Mexico and southern China. In general the species are usually evergreen shrubs or small trees with foliage of leathery consistency. They occur in both the mixed mesophytic and evergreen broad-leafed forest formations of China, where they tend to be understory plants. *S. tinctoria* (L.) L'Her. the sweetleaf of the southern Appalachians and adjacent piedmont, grows in moist situations in company with *Magnolia, Persea, Liquidambar, Platanus, Prunus, Ilex, Cornus, Nyssa* and other woody plants typical of the region. It seems never to comprise more than 1 percent of the total tree and shrub individuals. The species of *Symplocos* are, in general, plants of humid climates where marked dry seasons do not occur. However *Symplocos prionophylla* Hemsley is a common small tree of the bosque mesofilo de montaña between altitudes of 2500 and 6000 feet in Nayarit and western Jalisco, Mexico. In these localities it tends to grow in protected canyons, but experiences a winter dry season of 4 to 6 months.

Symplocos appears as one of the dominants in the low ground Paleogene and early Neogene floras around the northern hemisphere in middle latitudes.

Occurrence.—Kisinger Lakes, Tipperary. Type PA 5681; paratypes PA 5630, PA 5671, PA 5724; topotypes PA 5731, PA 5732.

Family ASPIDIACEAE

Genus *Thelypteris* Schmiedal

Thelypteris (Cyclosorus) iddingsi (Knowlton), new combination

(Pl. 1, figs. 1, 2)

Asplenium iddingsi Knowlton, U.S. Geol. Surv. Mon. 32, part 2, p. 666, pl. 80, figs. 9, 10, pl. 81, fig. 1, 1899.

Dryopteris weedii Knowlton, Berry, U.S. Geol. Surv. Prof. Paper 165, p. 63, pl. 7, fig. 1, 1930b.

Berry illustrated the common fern of the lower Middle Eocene in the Wind River basin correctly, but mistakenly identified it with *Dryopteris weedii* Knowlton (1930*b*: pl. 80, fig. 8; pl. 81, fig. 2) which is clearly a different entity. Knowlton made a curious error in assigning the fern to *Asplenium* from which it differs in many respects. The fossil fern bore large leaves, perhaps as much as two feet in length judging from the material available. The foliage was probably twice pinnate. The pinnules are united for about half their length and bear round sori, dorsal on the veins. There are 6–7 simple, slightly curving veins. The lower pair only, unite at the juncture of the pinnules. According to Copeland (1947:140–143) "only the lowest veins must unite to indicate the place of a species in *Cyclosorus*." The illustration shows the characteristic venation. At present the consensus of pteriodologists seems to be that *Cyclosorus* is a subgenus of *Thelypteris*. Therefore the genus *Thelypteris* is used for these ferns rather than *Cyclosorus*. The closest correspondences of the fossils are with living species annotated *Cyclosorus*, such as *C. dentatus* Forskal and *C. euryphyllus* Ros. (*C. evoluta* C. Christensen) of the tropical Pacific and India. There have been between 200 and 300 species of *Cyclosorus* recognized in tropical and subtropical regions of the world. They are most abundant in the Oriental tropics. Five species have been reported from South Africa and two from New Zealand. *C. dentatus* has been found in tropical Mexico also.

The taxonomy of the ferns offers great difficulties—it still seems to be in a somewhat chaotic state. *Cyclosorus* in the past has been assigned variously to *Aspidium*, *Dryopteris*, *Thelypteris* and probably *Phegopteris*. It may be said also that if the taxonomy of the living ferns may seem chaotic in certain areas, that of the fossil ferns clearly deserves that term.

Relationships to other fossil species must remain somewhat doubtful for *T. iddingsi* due to the inadequacy of illustrations and descriptions. *Goniopteris claiborniana* Berry (1924:44, pl. 5, pl. 6) undoubtedly represents the same genus but a different species. Berry's comments are worth reading for anyone working on the taxonomy of ferns, living or fossil. His "*Goniopteris*" fossils are common in the Yegua and Columbia Formations (Middle Eocene) of the eastern Gulf Coast. Frequency, common.

Occurrence.—Kisinger Lakes, Tipperary, Yellstone "basic breccia." Hypotype PA 5600.

Thelypteris (Cyclosorus) weedii, new combination

Dryopteris weedii Knowlton, U.S. Geol. Surv. Mon. 32:2, pp. 669–670, pl. 80, fig. 8; pl. 81, fig. 2, 1899.

This species is represented by fragments of pinnae. The genus is more properly designated as *Thelypteris* (or *Cyclosorus*). It differs from *T. iddingsi* in having from 18–20 lateral veins instead of 6–7, and in the deeper dissection of the pinnules which are separate to within 4 to 5 mm. from the rachis. The original material was from the "early basic breccia" of Yellowstone Park. Knowlton's figures of this fossil fern appear to be incorrectly drawn. The basal veins in each pinnule unite to form one vein which rises vertically for approximately 4 mm. to the juncture of the pinnules, which are united in the lower 5 mm. The species cited by Berry as *Dryopteris weedii* Knowlton is not the same as that illustrated in Professional Paper 165, pl. 7, fig. 1. This is *T. iddingsi* (Knowlton) MacGinitie. A closely similar living species *Thelypteris totta* (Thunb.) Schelpe, may be found growing abundantly with *Acrostichum* around the borders of the mangrove swamps along the tropical west coast of Mexico. Frequency, rare.

Occurrence.—Tipperary. Homeotype PA 5601.

Family ULMACEAE
Genus *Zelkova* Spach
Zelkova nervosa (Newberry) Brown
(Pl. 2, fig. 2; pl. 9, fig. 3)

Zelkova nervosa (Newberry) Brown, Wash. Acad. Sciences Jour., vol. 36, p. 346, 1946.
Sapindus dentoni Lesquereux, Berry, U.S. Geol. Surv. Prof. Paper 165, p. 71, 1930.

The rather uncommon leaf impressions of *Zelkova* are assigned to the Green River species. The leaves of this genus are difficult to separate into distinct species because of the marked range in size at any particular locality and the relatively few diagnostic characters. There seems to be a general increase in size from the Early Eocene to the later Tertiary. The largest leaf found at the Kisinger Lakes locality is 5 cm. in length and 2 cm. in width. The oldest in the series is from the Lost Cabinian flora of the Wind River basin. Both the Kisinger Lakes forms and those of the Wind River beds have a slightly higher density of secondaries than those from Green River or Florissant. The material called *Sapindus dentoni* by Berry undoubtedly represents this species. See the note under *Sapindus* in this paper. No definite fossils of *Sapindus* have so far been found in the flora.

Occurrence.—Kisinger Lakes, Tipperary. Hypotypes PA 5603, PA 5631; topotype PA 5632.

LITERATURE CITED

AXELROD, D. I.
  1966. Origin of deciduous and evergreen habits in temperate forests. Evolution 20(1):1–15.
BAILEY, E. W., and I. W. SINNOTT
  1915. Investigations on the phylogeny of the angiosperms, 5: foliar evidences as to the ancestry and early climatic environment of the angiosperms. Amer. Jour. Bot. 2(1):1–22.
  1916. Investigations on the phylogeny of the angiosperms, 6: climatic distribution of certain types of angiosperm leaves. Amer. Jour. Bot. 3(1):23–39.
BANDY, O. L., and J. C. INGLE, JR.
  1970. Neogene planktonic events and radiometric scale, California. Geol. Soc. Amer., special paper 124:131–172.
BECKER, H. F.
  1960. The Tertiary Mormon Creek flora from the upper Ruby River basin in southwestern Montana. Paleontographica, 107, Abt.B.
BELL, W. A.
  1949. Upper Cretaceous and Paleocene floras of western Alberta. Canada Dept. Mines and Resources, Geol. Surv. Bull. 13.
BERGGREN, W. A.
  1969. Cenozoic chronostratigraphy, planktonic foraminiferal zonation and the radio-metric time scale. Nature 224:1072–1075.
BERRY, E. W.
  1914. Upper Cretaceous and Eocene floras of South Carolina and Georgia. U.S. Geol. Surv. Prof. Paper 84.
  1916. The Lower Eocene floras of southeastern North America. U.S. Geol. Surv. Prof. Paper 91.
  1924. The Middle and Upper Eocene floras of southeastern North America. U.S. Geol. Surv. Prof. Paper 92.
  1930a. Revision of the Lower Eocene Wilcox flora of the southeastern states, U.S. Geol. Surv. Prof. Paper 156.
  1930b. A flora of Green River age in the Wind River Basin of Wyoming. U.S. Geol. Surv. Prof. Paper 165:55–81.
BRADLEY, W. H.
  1961. Geologic and geophysical studies in parts of Wyoming, southeastern Idaho and northeastern Utah. U.S. Geol. Surv. Prof. Paper 424A:25.
BRADLEY, W. H., and H. P. EUGSTER
  1969. Geochemistry and paleolimnology of the trona deposits and associated authigenic minerals of the Green River Formation of Wyoming. U.S. Geol. Surv. Prof. Paper 496B: 21–25.
BROWN, R. W.
  1934. The recognizable species of the Green River flora. U.S. Geol. Surv. Prof. Paper 185C.
  1939. Some American fossil plants belonging to the Isoetales. Wash. Acad. Sciences Jour. 29(6):261–269.
  1943. A climbing fern from the Upper Cretaceous of Wyoming. Wash. Acad. Sciences Jour. 33: 141–142.
  1944. Temperate species in the Eocene flora of the southeastern United States. Wash. Acad. Sciences Jour. 34(11):349–351.
  1946. Alterations in some fossil and living floras. Wash. Acad. Sciences Jour. 36:344–355.
  1962. Paleocene flora of the Rocky Mountains and Great Plains. U.S. Geol. Surv. Prof. Paper 375.
BRUZON, E., and P. CARTON
  1930. Le climat de l'Indochine. Exposition Coloniale International, Paris 1931. Observatoire Central de l'Indochine Francaise.
CHANDLER, M. E. J.
  1960, 1961, 1964. The lower Tertiary floras of southern England. British Museum, 5 vols.
CHANEY, R. W., and E. J. SANBORN
  1933. The Goshen flora of west-central Oregon. Carnegie Inst. Wash. Publ. 439.

COCKERELL, T. D. A.
　1925. Plant and insect fossils from the Green River Eocene of Colorado. U.S. Nat. Mus. Proc. 66:1–13.

COPELAND, E. B.
　1947. Genera Filicum. Annales Cryptogamici et Phytopathologici. Chronica Botanica Co.

CULBERTSON, W. C.
　1962. Laney Shale Member and Tower Sandstone Lentil of the Green River Formation, Green River area, Wyoming. U.S. Geol. Survey Prof. Paper 450-C, pp. 54–56.

DAWSON, J. W.
　1875. Rept. geology and resources of the region in the vicinity of the 49th parallel. British North American boundary commission. App. A:327–331.

DEVEREUX, L.
　1967. Oxygen isotope paleotemperature measurements on New Zealand Tertiary fossils. New Zealand Jour. Sci. 10:988–1011.

DILCHER, D.
　1971. The Eocene Green River flora of northwestern Colorado and northeastern Utah, Mac-Ginitie, H. D. Review. Jour. Paleon. 45:739–740.

DILCHER, D., and G. DOLPH
　1970. Fossil leaves of *Dendropanax* from Eocene sediments of southeastern North America. Amer. Jour. Sci. 57:153–160.

DORF, E.
　1938. Stratigraphy and paleontology of the Fox Hills and Medicine Bow Formations of southern Wyoming and northwestern Colorado. Carnegie Inst. Wash. Publ. 508:1–78.
　1942. Flora of the Lance Formation at its type locality, Niobrara County, Wyoming. Carnegie Inst. Wash. Publ. 508:79–159.
　1960. Tertiary fossil forests of Yellowstone National Park. Billings Geol. Soc. 11th Ann. Field Conf., 253–260.
　1969. Paleobotanical evidence of Mesozoic and Cenozoic climatic changes. Proc. North Amer. Paleon. Conv. Proc., Part D:323–346.

EVERNDEN, J. F., D. E. SAVAGE, G. H. CURTIS, and G. T. JAMES
　1964. Potassium-argon dates and the Cenozoic mammalian chronology of North America. Amer. Jour. Sci. 262:145–198.

GERMERAAD, J. H., C. A. HOPPING, and J. MULLER
　1968. Palynology of Tertiary sediments from tropical areas. Rev. Paleobotany and Palynology, 6(¾):189–348.

GRAY, JANE
　1960. Temperate pollen genera in the Eocene (Claiborne) flora, Alabama. Science 132(3430): 808–810.
　1969. Preparation Techniques for Pollen Analysis Part III *in* Kummel, Bernhard, and Raup, David. Preparation Techniques in Paleontology. San Francisco: W. H. Freeman, 323 pp.

HADEN-GUEST, WRIGHT, and TECLAFF, eds.
　1956. A world geography of forest resources. Ronald Press.

HARPER, R. M.
　1928. Economic botany of Alabama. Catalogue of the trees, shrubs, and vines of Alabama. Geol. Surv. Alabama, Mon. 9, pt. 2. University of Alabama.

HAY, R. L.
　1956. Pitchfork Formation, detrital facies of early basic breccia, Absaroka Range, Wyoming. Amer. Assoc. Petr. Geol. Bull. 40(8):1863–1898.

HEER, O.
　1855. Flora Tertiaria Helvetiae. Winterthup.

HOLLICK, A.
　1936. The Tertiary floras of Alaska. U.S. Geol. Surv. Prof. Paper 182.

HUZIOKA, K., and E. TAKAHASI
　1970. The Eocene flora of the Ube coal-field, southwest Honshu, Japan. Jour. Mining College of Akita University, ser. A, 4(5):1–88.

JENKINS, D. G.
  1968. Planktonic foraminifera as indicators of New Zealand paleotemperatures. Tuatara 16: 32–37.
KEEFER, W. R.
  1957. Geology of the DuNoir area, Fremont County, Wyoming. U.S. Geol. Surv. Prof. Paper 294E:155–221.
  1965a. Stratigraphy and geologic history of the uppermost Cretaceous, Paleocene, and Lower Eocene rocks in the Wind River basin, Wyoming. U.S. Geol. Surv. Prof. Paper 495A.
  1965b. Geologic history of the Wind River basin. Bull. Amer. Assoc. Petr. Geol. 49(11).
  1969. Geology of petroleum in the Wind River basin, central Wyoming. Bull. Amer. Assoc. Petr. Geol. 53(9):1839–1865.
  1970. Structural geology of the Wind River basin. U.S. Geol. Surv. Prof. Paper 495D.
KIRA, T.
  1948. On the classification of altitudinal and climatic zones by the warmth index. Kanchi-Nogaku (Boreal Agriculture) 2(2):143–173.
  1949. Forest zones of Japan. Forestry Explanat. Ser. 17. Jap. Forest. Tech. Assoc. Tokyo.
KLUCKING, E. P.
  1962. An Oligocene flora from the western Cascades, with an analysis of leaf structure. Ph.D. thesis, Univ. Calif. Berkeley.
KNOWLTON, F. H.
  1899. Fossil flora of the Yellowstone National Park. U.S. Geol. Surv. Mon. 32:651–791.
  1923. Revision of the flora of the Green River Formation. U.S. Geol. Surv. Prof. Paper 131: 133–182.
KÖPPEN, W.
  1931. Grundriss der Klimakunde. Leipzig: Walter de Gruter Co.
KÖPPEN-GEIGER
  1954. Klima der Erde. Darmstadt: Justus Perthes.
KOSTERMANS, A. J.
  1962. The Asiatic species of *Persea* Miller. Reinwardtia 6:189–194.
LA MOTTE, R. S.
  1952. Catalogue of Cenozoic plants of North America through 1950. Geol. Soc. Amer. Mem. 51.
LARSEN, J. A.
  1930. Forest types of the northern Rocky Mountains and their climatic controls. Ecology 11(4):631–672.
LEOPOLD, E. B., and H. D. MACGINITIE
  1972. Development and Affinities of Tertiary floras in the Rocky Mountains, *in* A. Graham, ed., "Floristics and Palaeofloristics of Asia and Eastern North America." Amsterdam: Elsevier Press.
LESQUEREUX, L.
  1878. The Tertiary flora. U.S. Geol. Surv. Terr. Rept., 7.
  1883. The Cretaceous and Tertiary flora. U.S. Geol. Surv. Terr. Rept., 8.
LI, H. L.
  1963. Woody flora of Taiwan. Morris Arb. Mon. University of Pennsylvania, Livingston Publ. Co.
LOHMAN, K. E., and G. W. ANDREWS
  1968. Late Eocene nonmarine diatoms from the Beaver Divide Area Fremont County, Wyoming: U.S. Prof. Paper 593-E, p. E1-E26.
LOVE, J. D.
  1939. Geology along the southern margin of the Absaroka Range, Wyoming. Geol. Soc. Amer. Spec. Paper 20.
  1970. Cenozoic geology of the Granite Mountains area, central Wyoming. U.S. Geol. Surv. Prof. Paper. 495.
MACGINITIE, H. D.
  1941. A Middle Eocene flora from the central Sierra Nevada. Carnegie Inst. Wash. Publ. 534.
  1955. Fossil plants of the Florissant beds, Colorado. Carnegie Inst. Wash. Publ. 599.

1962. The Kilgore flora, a late Miocene flora from northern Nebraska. Univ. Calif. Publ. Geol. Sci. 35(2).

1969. The Eocene Green River flora of northwestern Colorado and northeastern Utah. Univ. Calif. Publ. Geol. Sci. 83.

MAI, D. H.
1964. Die mastixioiden floren im Tertiär der Oberlausitz. Paläeontologische Abhandlungen, Band 11, Heft 1. Berlin: Akademis-Verlag.

MASON, H. L.
1947. Evolution of certain floristic associations in Western North America. Ecological Monographs 17:202–210.

MOORE, R., and B. RATCLIFFE
1971. The record in the rocks. Audubon 73(1):17.

MORA, C., and E. JÁUREGUI
1965. Isotermas extrema e indice de aridez en la Republica Mexicana. Universidad Nacional Autonoma de Mexico. Instituto de Geografia.

NELSON, W. H., and W. G. PIERCE
1968. Wapiti Formation and Trout Creek Trachyandesite, northwestern Wyoming. U.S. Geol. Surv. Bull. 1254-H:H1–H11.

NEMEJč, F.
1964. Biostratigraphic sequence of floras in the Tertiary of Czechoslovakia. Casopsis Mineral, Geol. 9.

NEWBERRY, J. S.
1898. The later extinct floras of North America. U.S. Geol. Surv. Mon. 35.

PENNINGTON, T. D., and J. SARUKHAN
1968. Arboles tropicales de Mexico. Commonwealth Forestry Institute, University of Oxford.

POKROVSKAIA, I. M., Ed.
1966. Paleopalynology. Ministry of Geology of the U.S.S.R. Scientific Institute of Geological Sciences Investigations. Issue 141, 2.

POTBURY, S. S.
1935. The La Porte flora of Plumas County, California. Carnegie Inst. Wash. Publ. 465-11.

REID, E. M., and M. E. J. CHANDLER
1933. The London Clay flora. London: British Mus. Nat. Hist.

ROEHLER, H. W.
(In press). Stratigraphy of the Washakie Formation in the Washakie Basin, Wyoming: U.S. Geol. Survey Bull. 1369.

ROHRER, W. L.
1966. Geologic quadrangle map, Kisinger Lakes Quadrangle, Fremont County, Wyoming. U.S. Geol. Surv. Geol. Quad. Map GQ-724.

1968. Geologic map of the Fish Lake Quadrangle, Fremont County, Wyoming. U.S. Geol. Surv. Geol. Quad. Map GQ-724.

1969. Preliminary geologic map of the Sheridan Pass quadrangle, Fremont and Teton Counties, Wyoming. U.S. Geol. Survey open-file map.

ROHRER, W. L., and J. D. OBRADOVICH
1969. Age and stratigraphic relations of the Tepee Trail and Wiggins Formations, northwestern Wyoming. U.S. Geol. Surv. Prof. Paper 650B:57–62.

ROUSE, J. T.
1935. The volcanic rocks of the Valley area, Park County, Wyoming. Amer. Geophys. Union Trans., 16th Ann. Meeting, pt. 1:274–284. Nat. Research Council.

RZEDOWSKI, J., and R. McVAUGH
1966. La vegetacion de Nueva Galicia. Contri. Univ. Mich. Herbarium 9(1).

SAUER, J.
1964. Revision of *Canavalia*. Brittonia, 16:106–181.

SEIN, MA KHIN
1961. Palynology of the London Clay. Ph.D. thesis, Univ. College, London.

SMITH, C. E.
1960. A revision of *Cedrela*. Fieldiana: Botany, 29(5):295–341.

SOCIETY OF AMERICAN FORESTERS
 1954. Report of the Committee on forest types.
ST. JOHN, O. H.
 1883. Report on the geology of the Wind River district, in Hayden, F. V. 12th Ann. Rept., U.S. Geol. and Geog. Surv. Terr., pt. 1, 1878:173–269.
TANAI, T.
 1970. The Oligocene floras from the Kushiro coal field, Hokkaido, Japan. Faculty of Sci. Jour., Hokkaido Univ.; series IV, Geol. and Min. 14(4):383–514.
THORNTHWAITE, C. W.
 1948. An approach to a rational classification of climate. Geog. Rev. 38:55–94.
TSCHUDY, R. H., and R. A. SCOTT, eds.
 1969. Aspects of palynology. New York: Wiley-Interscience Pub.
VAN HOUTEN, F. B.
 1964. Tertiary geology of the Beaver River Area Fremont and Natrona Counties, Wyoming: U.S. Geol. Survey Bull. 1164, 99 pp.
WANG, CHI-WU
 1961. The forests of China. Maria Moors Cabot Found. Publ. 5. Harvard Univ. Press.
WEBB, L. J.
 1959. Physiognomic classification of Australian rain forests. Journ. Ecology 47:551–570.
WEST, R. M.
 1969. Geology and vertebrate paleontology of the northeastern Green River Basin, Wyoming: Wyoming Geol. Assn. Guidebook, 21st Field Conf., pp. 77–92.
WILSON, W. H.
 1963. Correlation of volcanic rock units in the southern Absaroka Mountains, northwest Wyoming. Wyoming Univ. Contr. Geol. 2(1):13–20.
WODEHOUSE, R. P.
 1933. Tertiary Pollen-II The oil shales of the Eocene Green River Formation: Torry Bot. Club Bull. 60:479–524.
WOLFE, J. A.
 1966. Tertiary plants from the Cook Inlet region, Alaska. U.S. Geol. Surv. Prof. Paper 398B.
 1969. Paleogene floras from the Gulf of Alaska. Open file rept., U.S. Geol. Surv.
 1971. Tertiary climatic fluctuations and methods of analysis of Tertiary floras. Palaeogeography, Palaeoclimatology, Palaeoecology 9:27–57.

PLATE 1

Fig. 1. *Thelypteris iddingsi* (Knowlton) MacGinitie. Hypotype. U.C. PA 5600.

Fig. 2. The same, × 2.3 to show venation.

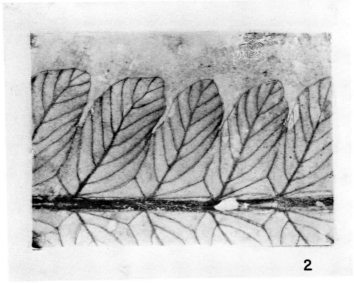

PLATE 2

Fig. 1. *Chamaecyparis* sp. Type. × 5, U.C. PA 5602.

Fig. 2. *Zelkova nervosa* (Newberry) Brown. Hypotype. U.C. PA 5603.

Fig. 3. *Prunus nota* MacGinitie. Type. × 2.6. U.C. PA 5602.

Fig. 4. *Castaneoides aequalita* MacGinitie. Type. U.C. PA 5604.

PLATE 3

Fig. 1. *Prunus nota* MacGinitie. Type. U.C. PA 5606.

Fig. 2. *Populus wyomingiana* (Berry) MacGinitie. Hypotype. U.C. PA 5607.

Fig. 3. The same, × 10 to show venation.

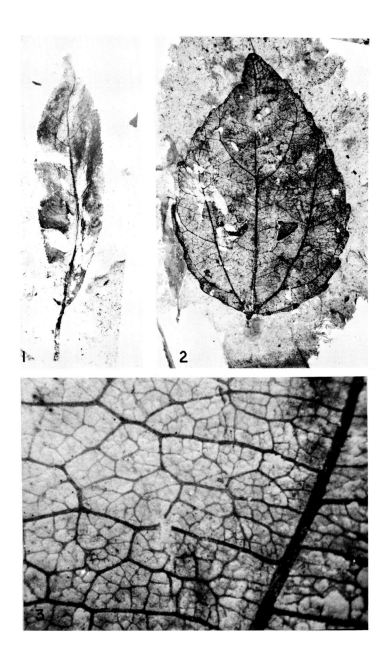

PLATE 4

Fig. 1. *Salix molesta* MacGinitie. Paratype. U.C. PA 5608.
Fig. 2. Portion of fig. 1, ×4, to show venation.
Fig. 3. *Salix molesta* MacGinitie. Type. U.C. PA 5609.

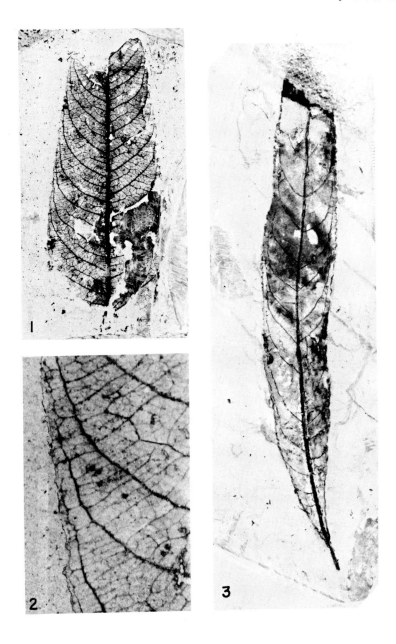

PLATE 6

Fig. 1. *Laurophyllum quotidiana* MacGinitie. Type. U.C. PA 5616.

Fig. 2. *Laurophyllum fremontensis* (Berry) MacGinitie. Hypotype. U.C. PA 5615.

Fig. 3. Portion of fig. 2, ×3 to show venation.

Fig. 4. *Spirodela magna* MacGinitie. Type. ×2, U.C. PA 5617.

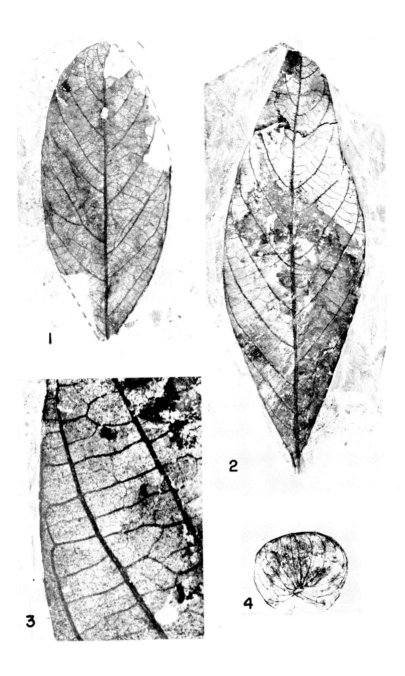

PLATE 7

Fig. 1. *Sparganium antiquum* (Newberry) Berry. Hypotype. ×4, U.C. PA 5624.

Fig. 2. Unidentified Lauraceae. Type. U.C. PA 5619.

Fig. 3. *Populus wyomingiana* (Berry) MacGinitie. Hypotype. U.C. PA 5622.

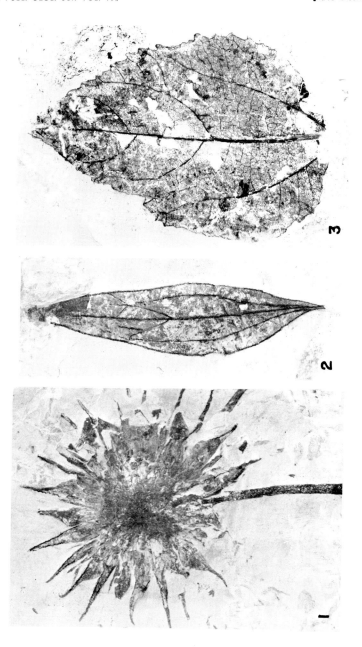

PLATE 8

Fig. 1. *Aristolochia solitaria* MacGinitie. Type. U.C. PA 5623.

Fig. 2. *Laurophyllum fremontensis* (Berry) MacGinitie. Hypotype. U.C. PA 5627.

Fig. 3. *Leguminosites mira* MacGinitie. Paratype. U.C. PA 5626.

Fig. 4. *Aristolochia solitaria* MacGinitie. Paratype. U.C. PA 5625.

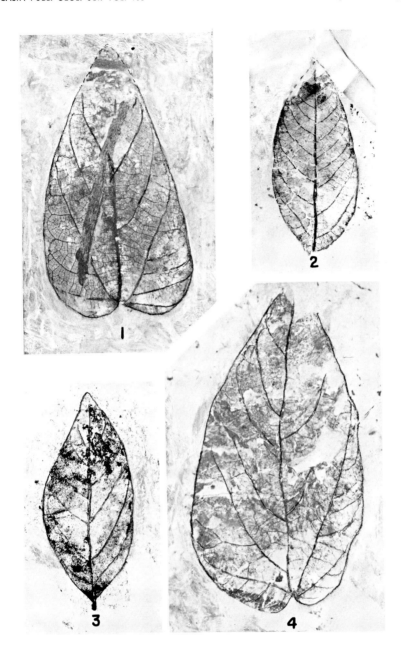

PLATE 9

Fig. 1. *Symplocos incondita* MacGinitie. Paratype. U.C. PA 5630.

Fig. 2. *Populus wyomingiana* (Berry) MacGinitie. Hypotype. ×2 to show venation. U.C. PA 5629.

Fig. 3. *Zelkova nervosa* (Newberry) Brown. Hypotype. ×2, U.C. PA 5631.

Fig. 4. *Populus wyomingiana* (Berry) MacGinitie. Hypotype. U.C. PA 5628.

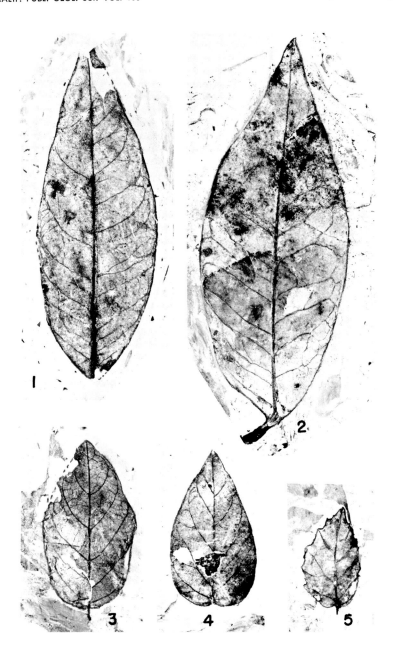

PLATE 11

Fig. 1. Unidentified leaf. U.C. PA 5639.

Fig. 2. *Leguminosites inlustris* MacGinitie. Type. ×1.8, U.C. PA 5638.

Fig. 3. *Leguminosites mira* MacGinitie. Paratype. U.C. PA 5640.

Fig. 4. Fig. 2, ×2.5 to show venation.

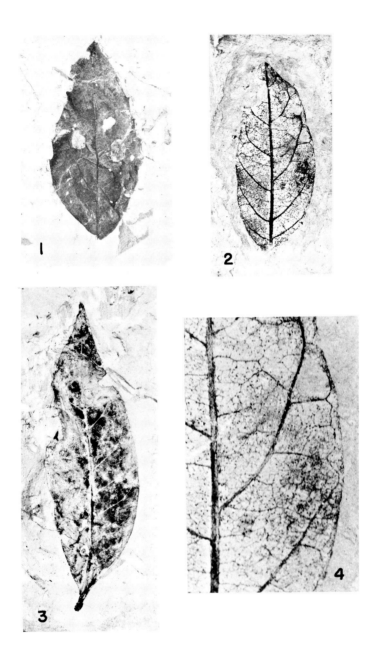

PLATE 12

Fig. 1. *Canavalia diuturna* MacGinitie. Paratype. U.C. PA 5642.

Fig. 2. *Laurophyllum quotidiana* MacGinitie. Paratype. U.C. PA 5641.

Fig. 3. *Canavalia diuturna* MacGinitie. Type. U.C. PA 5646.

Fig. 4. *Leguminosites wyomingensis* (Berry) MacGinitie. Hypotype. U.C. PA 5644.

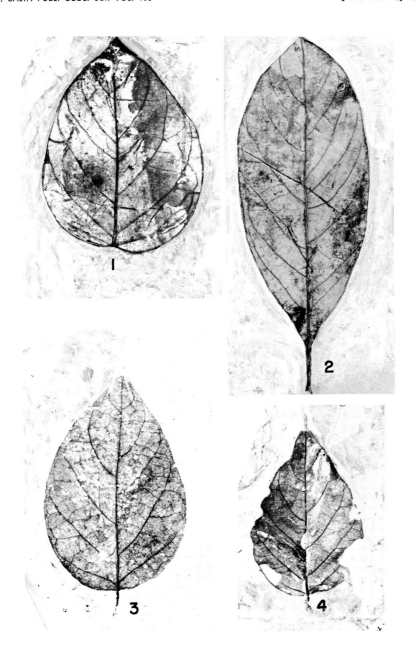

PLATE 13

Fig. 1. *Dendropanax latens* MacGinitie. Paratype. U.C. PA 5647.

Fig. 2. *Caesalpinites pecorae* (Brown) MacGinitie. Hypotype. PA 5648.

Fig. 3. *Luehea newberryana* (Knowlton) MacGinitie. Hypotype. ×2, U.C. PA 5650.

Fig. 4. Unidentified leaf. U.C. PA 5649.

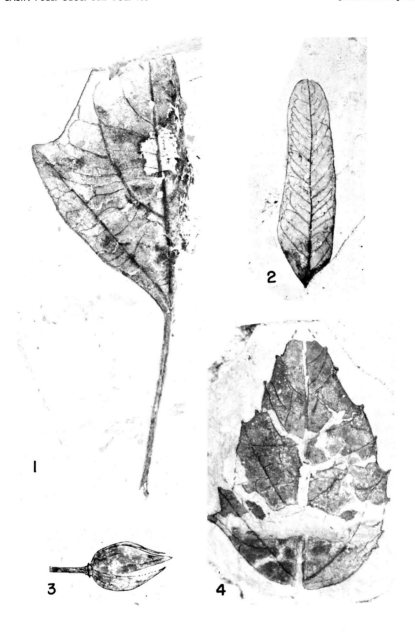

PLATE 14

Fig. 1. *Cissus* cf. *marginata* (Lesquereux) Brown. Hypotype. U.C. PA 5651.

Fig. 2. *Sparganium antiquum* (Newberry) Berry. Hypotype. ×3.5, U.C. PA 5652.

Fig. 3. Unidentified fruits. Type. ×2, U.C. PA 5653.

PLATE 15

Fig. 1. *Platanus browni* (Berry) MacGinitie. Hypotype. U.C. PA 5654.
Fig. 2. Unidentified fruit. Type. ×2, U.C. PA 5655.
Fig. 3. *Alafructus lineatulus* (Cockerell) MacGinitie. Hypotype. ×3.5, U.C. PA 5657.
Fig. 4. *Carpites araliodes* MacGinitie. Type. U.C. PA 5659.

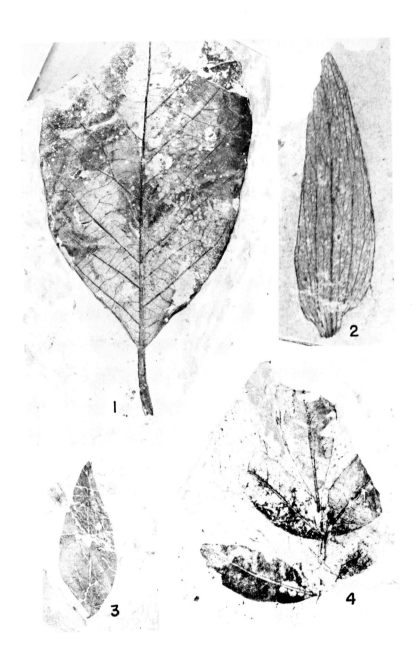

PLATE 19

Fig. 1. *Dendropanax latens* MacGinitie. Paratype U.C. PA 5679.

Fig. 2. *Dendropanax latens* MacGinitie. Paratype. U.C. PA 5680.

Fig. 3. *Acrostichum hesperium* Newberry. Hypotype. Fragment ×1.7 to show venation. U.C. PA 5677.

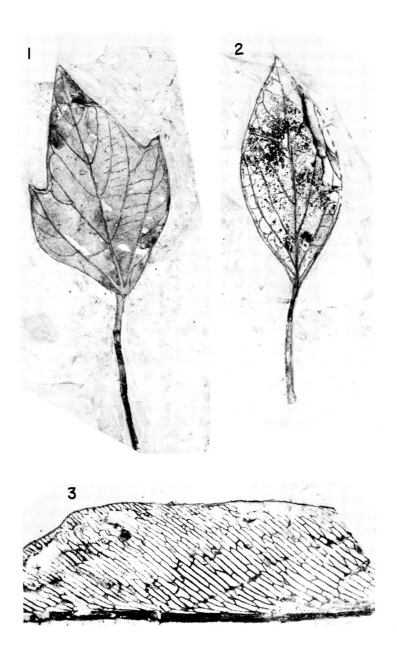

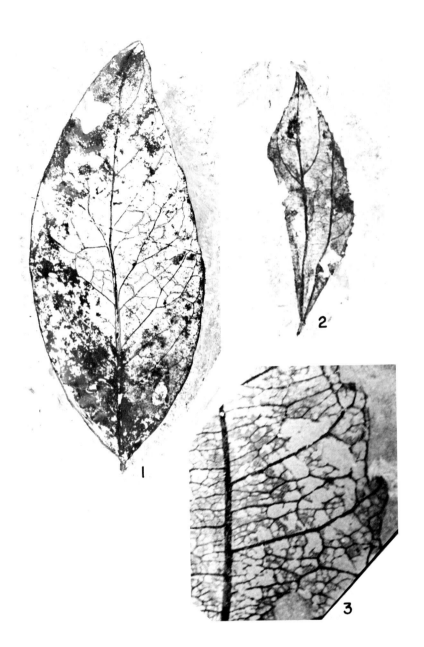

PLATE 21

Fig. 1. *Aleurites fremontensis* (Berry) MacGinitie. Hypotype. U.C. PA 5685.

Fig. 2. *Aleurites fremontensis* (Berry) MacGinitie. Hypotype. U.C. PA 5683.

Fig. 3. *Populus wyomingiana* (Berry) MacGinitie. Hypotype. U.C. PA 5684.

Fig. 4. *Spirodela magna* MacGinitie. Paratype. ×1.5, U.C. PA 5686.

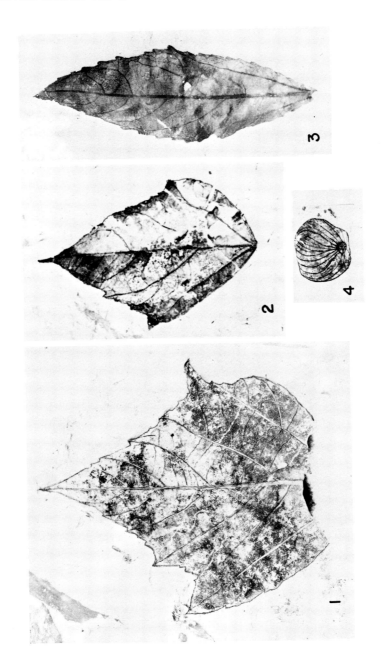

PLATE 22
Fig. 1. Portion of figure 2, ×5, to show venation.
Fig. 2. *Aleurites fremontensis* (Berry) MacGinitie. Hypotype. U.C. PA 5688.

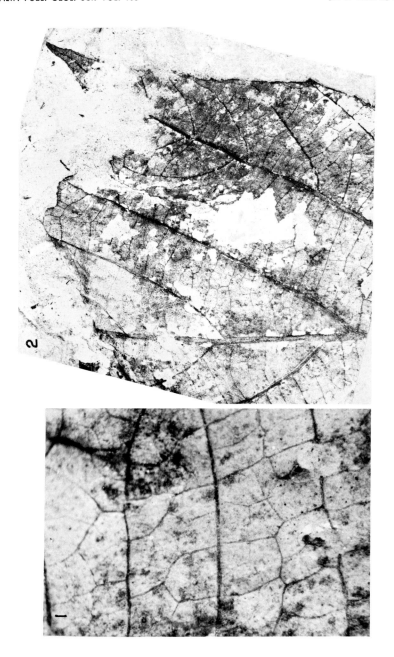

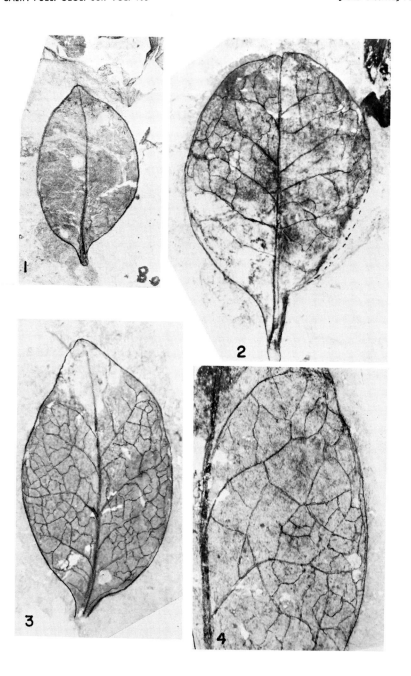

PLATE 24

Fig. 1. *Eugenia americana* (Knowlton) MacGinitie. Portion of fig. 2, ×7, to show venation.

Fig. 2. *Eugenia americana* (Knowlton) MacGinitie. Hypotype. ×1.7, U.C. PA 5694.

Fig. 3. *Cornus* sp. MacGinitie. Type. U.C. PA 5695.

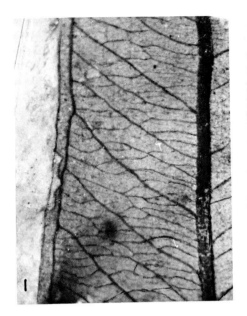

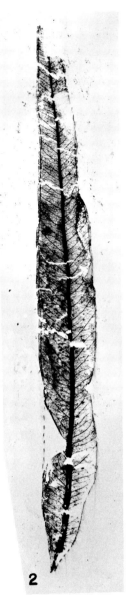

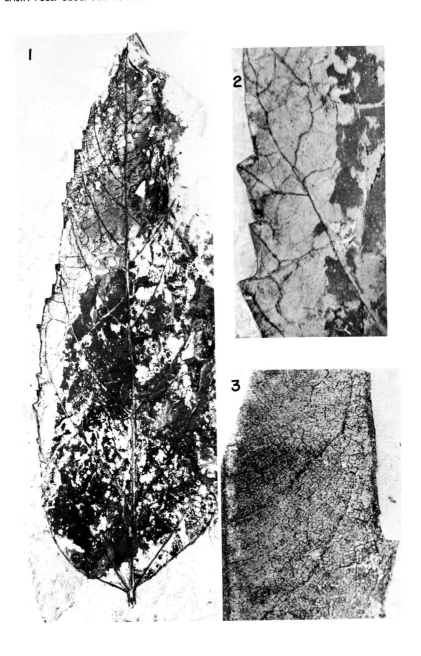

PLATE 27

Fig. 1. *Luehea newberryana* (Knowlton) MacGinitie. Hypotype. U.C. PA 5706.

Fig. 2. *Acalypha cirquensis* MacGinitie. Paratype. U.C. PA 5707.

PLATE 28

Fig. 1. *Schefflera insolita* MacGinitie. Paratype. U.C. PA 5712. Portion of a leaflet, ×5 to show venation.

Fig. 2. *Schefflera insolita* MacGinitie. Type. U.C. PA 5713.

Fig. 3. *Luehea newberryana* (Knowlton) MacGinitie. Hypotype. U.C. PA 5714.

Fig. 4. *Luehea newberryana* (Knowlton) MacGinitie. Hypotype. U.C. PA 5715.

PLATE 29

Fig. 1. *Dendropanax latens* MacGinitie. Type. U.C. PA 5716.
Fig. 2. Unidentified leaf. U.C. PA 5717.
Fig. 3. *Populus* seed capsules. U.C. PA 5718.

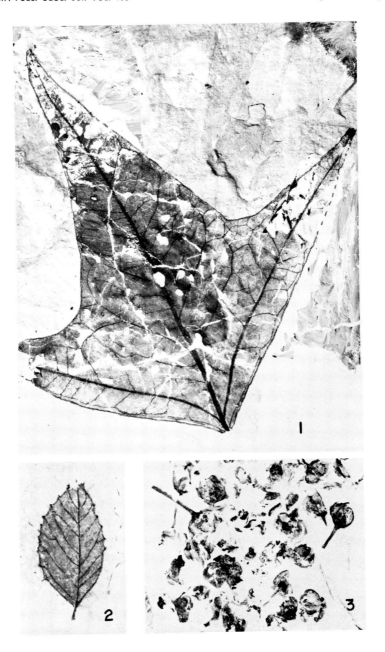

PLATE 30

Fig. 1. *Sterculia subtilis* MacGinitie. Portion of fig. 3, ×10, to show venation.

Fig. 2. *Sabalites florissanti* (Lesquereux) Berry. Hypotype. ×3, to show venation. U.C. PA 5721.

Fig. 3. *Sterculia subtilis* MacGinitie. Type. U.C. PA 5719.

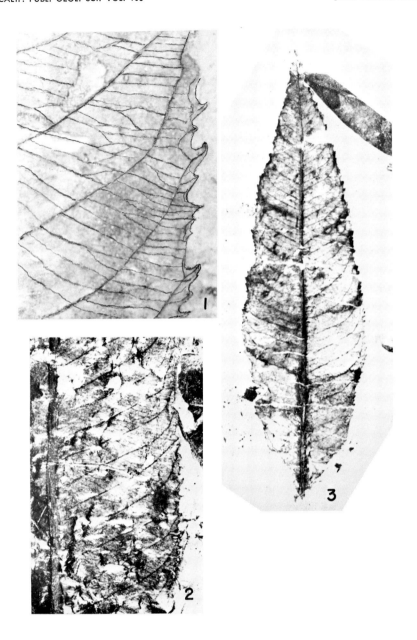

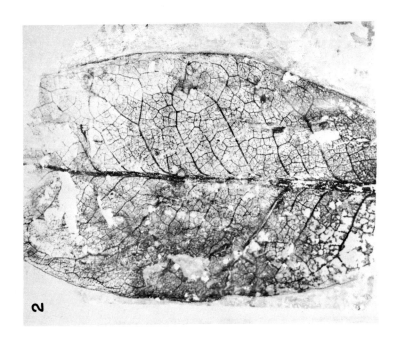

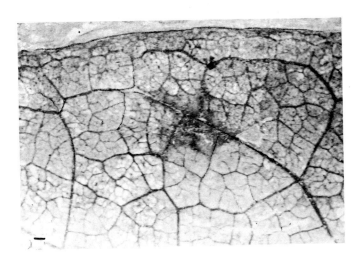

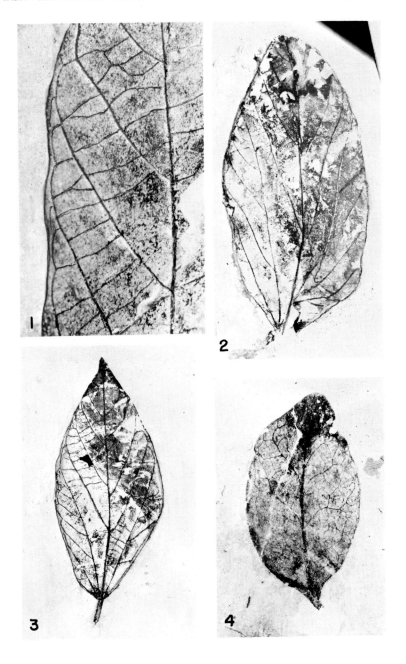

PLATE 34

Fig. 1. *Dipteronia wyomingense* (Berry) MacGinitie. Portion of pl. 25, fig. 4, ×6, to show venation. U.C. PA 5699.

Fig. 2. *Luehea newberryana* (Knowlton) MacGinitie. Portion of pl. 27, fig. 1, ×3, to show venation. U.C. PA 5706.

Fig. 3. *Canavalia diuturna* MacGinitie. Portion of pl. 12, fig. 3, ×2.2, to show venation. Paratype. U.C. PA 5730.

Fig. 4. *Platanus* fruits. U.C. PA 5729.

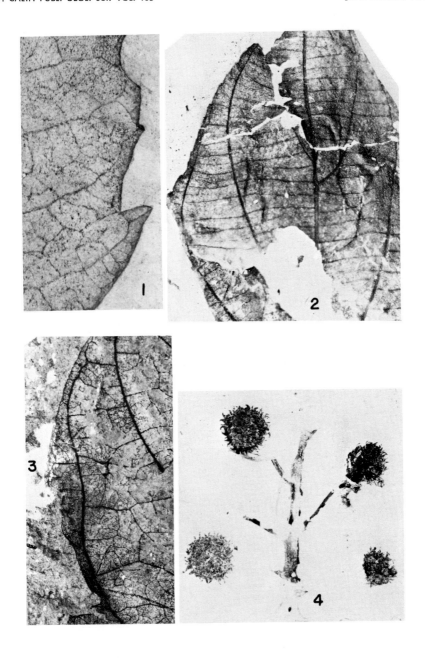

PLATE 35

Fig. 1. The Tipperary locality, looking north. The fossil bearing layer caps the hill.

Fig. 2. The original Kisinger Lakes locality. The layer containing fossil plants is at the top of the prominent outcrop in the left middle distance.

PLATE 36

Figs. 1–3. Monolete spores, Polypodiaceae or Aspidiaceae types.

        D3530(2)   88.3 × 2.0

        D3532B(3)   86.1 × 21.5

        D3532B(3)   106.9 × 3.1

Figs. 4–9, 11–14. Trilete spores undetermined.

        D3531A(3)   86.5 × 10.4

        D3530(2)   99.8 × 9.2

        D3531C(5)   81.2 × 21.7

        D3531C(5)   108.4 × 14.1

        D3532A(2)   87.1 × 4.3

        D3531A(1)   103.9 × 21.3

        D3532B(3)   96.8 × 18.9

        D3532B(2)   101.4 × 21.7

        D3532A(1)   105.4 × 15.6

        D3532B(2)   83.7 × 11.4

Fig. 10. *Lygodium* cf. *L. kaulfussi* Heer (Schizaeaceae).

        D3531A(3)   87.8 × 13.8

The second column in the captions for plates 36–45 shows the coordinate values for the grain on the particluar slide (see p. 50).

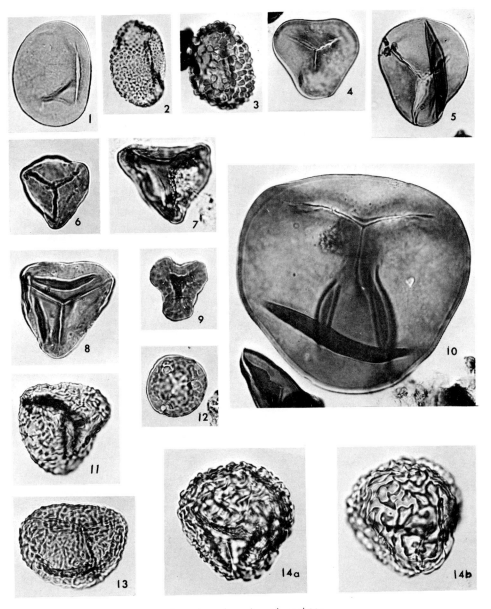

$\llcorner\lrcorner$ 50μ

PLATE 37

Figs. 1–3. Trilete spores, undetermined.

        D3532B(3)   84.2 × 21.2

        D3532A(2)   81.1 × 17.7

        D3532B(2)   101.7 × 8.5

Fig. 4. *Lycopodium* cf. *L. clavatum* (Lycopodiaceae).

        D3532B(3)   104.7 × 17.8

Fig. 5. *Osmunda* (Osmundaceae).

        D3532B(3)   83.2 × 13.4

Fig. 6. Trilete spore, undetermined.

        D3532B(2)   88.7 × 22.0

Fig. 7. cf. *Selaginella conduplicata* (Selaginellaceae?)

        D3532B(2)   95.2 × 14.5

Fig. 8. *Selaginella* cf. *S. densa* (Selaginellaceae).

        D3532B(3)   82.1 × 16.7

Fig. 9. cf. *Pteris* (Polypodiaceae).

        D3532B(2)   93.7 × 10.7

Fig. 10. Tetrad, trilete spores undetermined.

        D3531B(2)   88.3 × 19.0

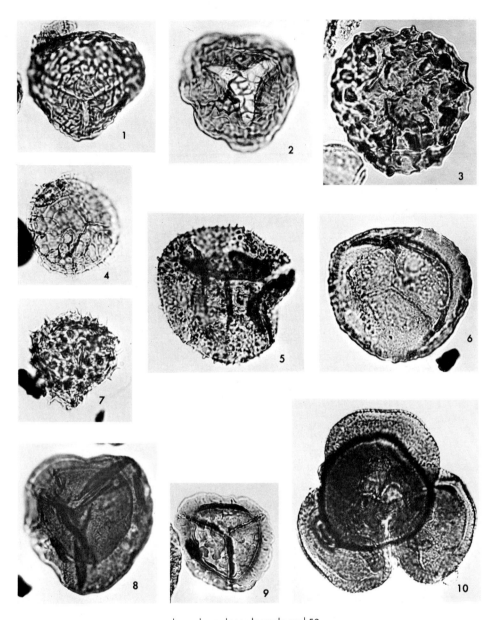

50 μ

PLATE 38

Fig. 1. Cf. *Cedrus* (Pinaceae).
D3532A(2) 106.0× 8.1

Fig. 2a, b. cf. *Pinus* (Pinaceae).
D3530(2) 98.9× 11.9

Fig. 3. Cf. *Keteleeria* (Pinaceae).
D3532A(2) 87.4× 12.8

Fig. 4. *Picea* (Pinaceae).
D3530(3) 95.6× 15.0

Figs. 5a, b, 6, 8. *Pinus* (Pinaceae).
D3532A(2) 97.6× 18.3
D3532A(2) 98.6× 14.3
D3532B(3) 107.8× 12.9

Fig. 7. *Tsuga* (Pinaceae).
D3530(3) 101.9×11.8

Fig. 9. Cf. *Dacrydium* (Podocarpaceae?).
D3532B(2) 86.0× 13.2

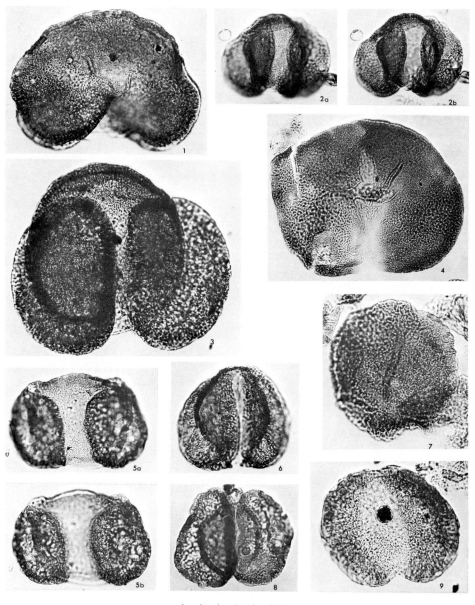

PLATE 39

Fig. 1. Taxodiaceae-Cupressaceae-Taxaceae type.
          D3531C(8)   86.1× 7.0
Fig. 2. *Glyptostrobus* or *Cryptomeria* (Taxodiaceae).
          D3532B(1)   102.4× 8.9
Fig. 3. Taxaceae or Taxodiaceae.
          D3530(3)   93.6× 11.5
Fig. 4. Cf. *Populus* (Salicaceae?).
          D3531A(3)   101.2× 4.3
Fig. 5, 6, 8, 9. Monosulcate pollen, cf. Palmae?
          D3531C(2)   98.3× 17.5
          D3531A(1)   94.8× 21.2
          D3531A(1)   100.8× 1.5
          D3531C(5)   100.5× 21.6
Fig. 7. Gingkoaceae or Cycadaceae.
          D3530(3)   82.3× 18.2
Fig. 10, 11, 12. Monosulcate, reticulate pollen, cf. Liliales or Amaryllidales.
          D3530(3)   94.0× 14.3
          D3531A(3)   100.0× 11.2
          D3531C(5)   99.9× 14.4
Fig. 13. *Eucommia* cf. *E. ulmoides* (Eucommiaceae).
          D3530(2)   86.3× 17.2
Fig. 14–22, 24, 25, 27, 28, 31. Tricolpate pollen undetermined.
          D3530(2)   93.1× 10.4
          D3530(2)   84.7× 3.8
          D3530(2)   108.2× 8.3
          D3530(3)   89.6× 15.2
          D3530(3)   108.4× 15.7
          D3531C(4)   101.5× 19.1
          D3530(3)   86.8× 11.2
          D3531C(8)   93.1× 21.1
          D3531C(5)   101.9× 13.0
          D3530(2)   107.5× 8.0
          D3530(2)   98.8× 7.2
          D3530(2)   105.3× 17.1
          D3531C(5)   105.9× 2.3
          D3531B(2)   107.0× 5.8
Figs. 23, 26, 29, 30a, b, 32, 33. *Tricolpites* (cf. *Platanus*, Platanaceae?).
          D3532B(3)   86.3× 12.8
          D3532B(3)   101.7× 11.7
          D3532B(3)   102.2× 20.0
          D3530(2)   107.7× 13.0
          D3530(2)   107.7× 13.0
          D3530(3)   97.0× 13.6
          D3532B(3)   93.5× 10.8

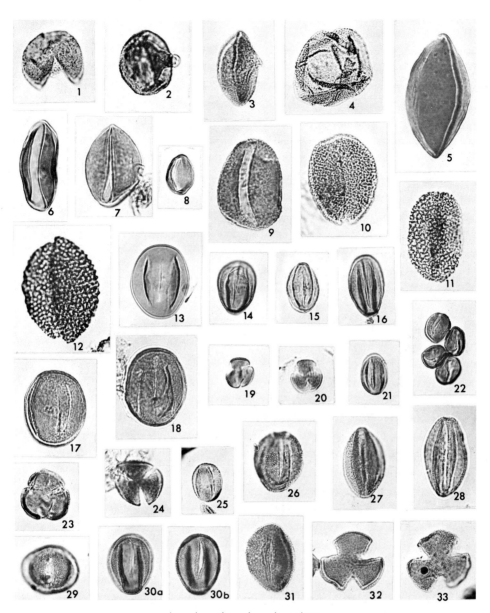

50μ

PLATE 40

Fig. 1, 3–7. Tricolpate pollen undetermined.

D3531C(5)	80.9 × 9.3
D3530(3)	105.2 × 9.8
D3530(2)	108.3 × 21.0
D3532B(3)	98.9 × 9.5
D3530(2)	89.3 × 17.5
D3530(2)	85.3 × 4.3

Fig. 2. Cf. *Prunus* (Rosaceae?).

D3531A(1)	94.0 × 10.1

Fig. 8. cf. *Gunnera* (Haloragidaceae?).

D3530(2)	102.5 × 13.7

Fig. 9. Tricolpate pollen undetermined.

D3530(3)	99.3 × 16.8

Fig. 10a, b. *Salix* (Salicaceae).

D3530(2)	109.8 × 11.2

Fig. 11–18. Tricolpate pollen, undetermined.

D3530(3)	98.9 × 15.0
D3531C(5)	92.9 × 5.9
D3530(3)	95.3 × 4.7
D3530(3)	108.4 × 11.8
D3530(2)	102.6 × 12.8
D3530(3)	100.0 × 10.8
D3531A(3)	84.6 × 5.2
D3531C(8)	96.8 × 16.7

Fig. 19–22. Rubiaceae?

D3532B(2)	81.5 × 2.5
D3531C(5)	93.1 × 2.0
D3530(3)	88.7 × 15.1
D3530(2)	101.3 × 12.1

Fig. 23, 24. Tricolpate pollen undetermined.

D3531C(8)	95.7 × 3.9
D3531C(8)	106.4 × 8.2

Fig. 25. *Pistillipollenites mcgregorii* Rouse.

D3531C(8)	89.3 × 11.2

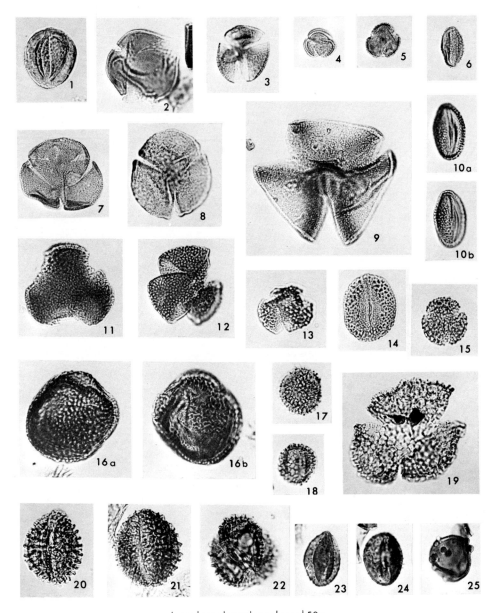

50μ

PLATE 41

Fig. 1. Cf. *Mucuna* (Leguminoseae).
        D3531C(5)   90.6 × 20.9

Fig. 2, 3. Cf. *Acer pseudoplatanus* (Aceraceae).
        D3530(3)   97.3 × 4.4
        D3530(3)   99.2 × 10.6

Fig. 4. Myrtaceae or Sapindaceae.
        D3530(3)   81.2 × 8.9

Fig. 5. cf. *Eugenia* (Myrtaceae?).
        D3531C(5)   109.4 × 14.9

Fig. 6–8. Tetracolpate pollen, undetermined.
        D3530(3)   94.3 × 16.7
        D3530(2)   99.7 × 9.8
        D3531C(8)   105.6 × 6.6

Fig. 9. Gramineae.
        D3530(3)   94.2 × 11.2

Fig. 10. Cf. Lemnaceae?
        D3530(3)   100.5 × 15.2

Fig. 11, 12. Monoporate pollen, undetermined.
        D3531C(5)   100.8 × 14.4
        D3530(3)   83.8 × 8.9

Fig. 13, 14. *Trema* (Ulmaceae).
        D3531C(5)   104.5 × 17.7
        D3531B(7)   97.4 × 19.9

Fig. 15–24. Triporate pollen undetermined.
        D3532B(1)   105.7 × 5.9
        D3530(3)   89.8 × 10.7
        D3532B(3)   98.8 × 16.3
        D3530(2)   94.7 × 19.0
        D3531C(8)   86.3 × 20.4
        D3531B(2)   93.0 × 3.0
        D3530(3)   91.4 × 11.6
        D3531A(1)   96.0 × 11.1
        D3531C(8)   95.5 × 5.7
        D3530(2)   94.6 × 15.8

Fig. 25a, b. Cf. *Carya* (Juglandaceae?).
        D3530(2)   98.9 × 7.3

Figs. 26, 27. *Carya* (Juglandaceae).
        D3530(3)   97.0 × 15.4
        D3531C(8)   84.3 × 12.2

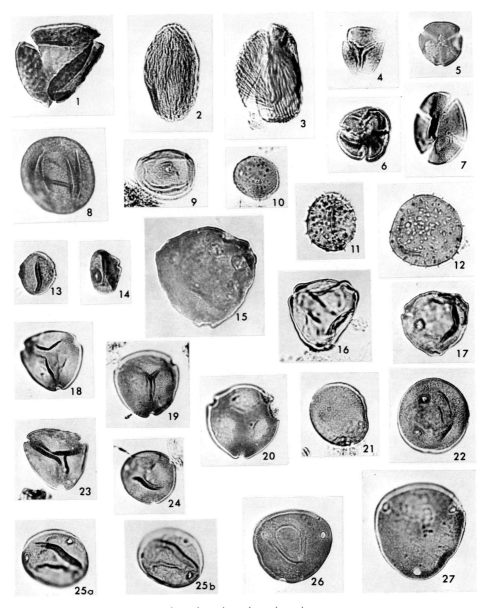

50μ

PLATE 42

Fig. 1. Cf. *Psoralea* (Leguminosae).
      D3531C(5)   84.0 × 5.6

Figs. 2, 3, 11. Cf. *Zelkova* (Ulmaceae).
      D3531A(3)   103.5 × 4.7
      D3531C(8)   87.1 × 15.3
      D3531A(3)   107.8 × 18.3

Figs. 4, 5. Triporate pollen, undetermined.
      D3531A(1)   87.2 × 6.8
      D3531A(1)   94.2 × 21.0

Figs. 6–8. Cf. *Schoutenia* (Tiliaceae?).
      D3531C(8)   95.7 × 4.7
      D3531C(2)   105.1 × 7.6
      D3530(2)   104.7 × 18.2

Fig. 9. Tetraporate pollen, undetermined.
      D3531A(1)   97.6 × 18.4

Figs. 10, 13. *Alnus* (Betulaceae).
      D3532B(3)   108.6 × 8.7
      D3532B(3)   104.6 × 20.2

Fig. 12. Tetraporate pollen, undetermined.
      D3532B(2)   84.9 × 4.2

Fig. 14. *Pterocarya* (Juglandaceae).
      D3532B(2)   102.2 × 5.8

Fig. 15. Cf. *Juglans* (Juglandaceae).
      D3531C(5)   96.6 × 2.0

Figs. 16a, b. Six-pored pollen, undetermined.
      D3530(2)   85.9 × 3.1

Figs. 17a, b. Chenopodiaceae or Amaranthaceae.
      D3532B(2)   106.0 × 18.1

Figs. 18–21, 23, 24, 26, 27. Tricolporate pollen, undetermined.
      D3532B(2)   106.8 × 10.0
      D3531C(5)   89.6 × 21.2
      D3531A(3)   91.8 × 4.7
      D3530(2)   90.8 × 5.0
      D3531A(3)   87.0 × 20.0
      D3531C(8)   87.3 × 13.4
      D3532B(3)   92.7 × 7.7
      D3531C(5)   103.9 × 4.5

Fig. 22. Compositae (contaminant).
      D3530(2)   101.1 × 9.0

Fig. 25. Cf. *Castanea* (Fagaceae).
      D3531C(8)   103.7 × 15.1

Fig. 28. Cf. *Serjania* (Sapindaceae?).
      D3532B(2)   99.3 × 11.1

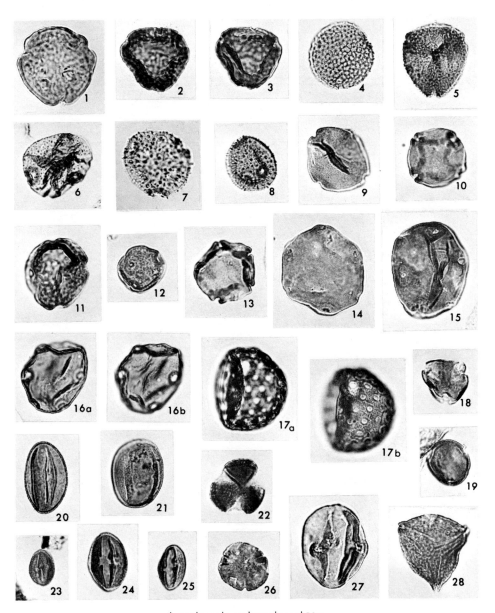

50 μ

PLATE 43

Fig. 1. Tricolporate pollen, undetermined.

    D3530(3)   90.0 × 16.0

Figs. 2, 3. *Dipteronia* cf. *D. sinensis* (Aceraceae).

    D3531C(8)   85.4 × 18.0

    D3532B(2)   86.7 × 6.0

Figs. 4, 5. *Acer* cf. *A. palmatum* (Aceraceae).

    D3531C(8)   90.8 × 11.1

    D3531C(8)   86.0 × 10.5

Figs. 6, 8. Cf. *Acer* (Aceraceae?).

    D3531A(1)   100.5 × 7.1

    D3532B(3)   102.7 × 9.0

Fig. 7. *Bauhinia* cf. *B. congesta* (Leguminoseae).

    D3532B(2)   109.2 × 6.8

Figs. 9–21. Tricolporate pollen undetermined.

    D3531A(3)   96.0 × 5.8

    D3531C(8)   103.6 × 16.9

    D3531C(5)   106.8 × 10.3

    D3531C(5)   98.2 × 9.2

    D3531C(2)   82.2 × 14.4

    D3531C(5)   109.6 × 15.5

    D3531C(8)   93.9 × 14.0

    D3530(2)   109.4 × 17.2

    D3531C(8)   98.9 × 5.2

    D3532B(3)   107.3 × 6.9

    D3530(3)   95.0 × 19.2

    D3531A(3)   103.4 × 17.7

    D3531B(7)   100.5 × 4.4

Figs. 22, 23. Cf. *Apeiba* (Tiliaceae).

    D3531C(8)   93.1 × 18.8

    D3531C(5)   97.3 × 7.8

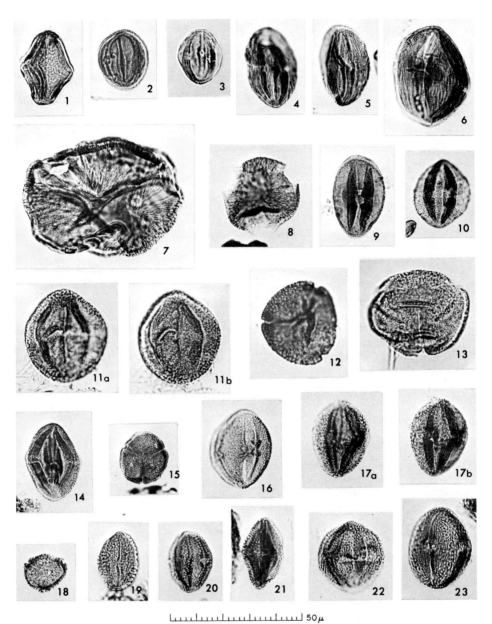

50μ

PLATE 44

Figs. 1, 2. *Triumfetta* (Tiliaceae).
        D3530(2)   87.9 × 5.4
        D3532B(3)   95.3 × 19.0
Figs. 3, 4. Tiliaceae?
        D3530(3)   82.3 × 13.8
        D3530(2)   106.5 × 8.8
Figs. 5–9. Tricolporate pollen, undetermined.
        D3531C(5)   102.8 × 2.7
        D3531C(8)   84.1 × 14.4
        D3531C(8)   86.8 × 3.6
        D3531C(8)   95.3 × 11.1
        D3532B(3)   107.4 × 11.7
Figs. 10a, b. Syncolporate pollen undetermined.
        D3531C(8)   97.1 × 9.2
Figs. 11, 13. *Ilex* (Aquifoliaceae).
        D3531C(8)   86.0 × 18.3
        D3531C(8)   92.6 × 10.2
Figs. 12a, b. *Luehea* (Tiliaceae).
        D3530(2)   85.8 × 7.2
        D3530(2)   85.8 × 7.2
Figs. 14, 16, 17, 18. Sterculiaceae? Section Helictereae.
        D3530(3)   86.0 × 9.0
        D3530(3)   99.3 × 19.2
        D3531A(1)   90.4 × 18.7
        D3531C(8)   96.7 × 21.1
Fig. 15. *Tilia* (Tiliaceae).
        D3530(3)   107.6 × 10.5

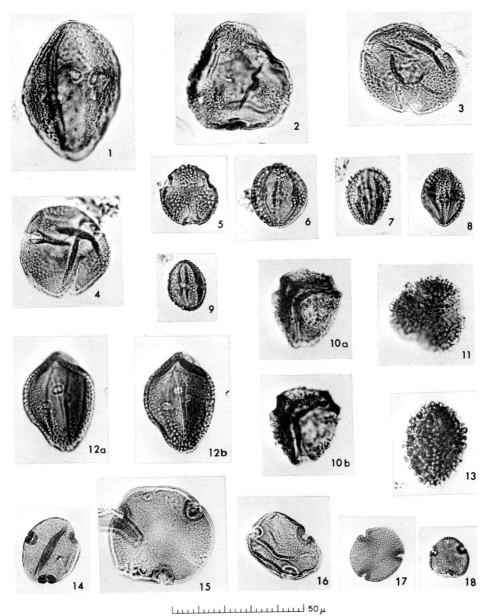

50 μ

PLATE 45

Figs. 1–3. Bombacaceae?

    D3531A(3)   101.5 × 16.2
    D3531A(1)   95.0 × 15.5
    D3531A(3)   102.7 × 9.6

Figs. 4, 5. *Bombax* (Bombacaceae).

    D3531C(5)   99.1 × 17.5
    D3531A(3)   87.1 × 4.0

Figs. 6–8. Brevicolporate pollen, undetermined.

    D3531A(3)   101.2 × 15.4
    D3531C(1)   82.0 × 10.2
    D3531B(2)   93.2 × 3.0

Fig. 9. Tetracolporate pollen, undetermined.

    D3531C(1)   90.3 × 20.0

Fig. 10. *Bernoullia* (Bombacaceae).

    D3530(3)   105.4 × 17.0

Figs. 11, 13, 14. Sapotaceae.

    D3531C(5)   107.1 × 17.5
    D3532B(2)   82.2 × 6.8
    D3531C(8)   90.7 × 15.4

Fig. 12. *Cedrela* cf. *C. mexicana* (Meliaceae).

    D3531C(5)   100.1 × 7.1

Fig. 15. Cf. *Equisetum* (Equisetaceae?).

    D3531A(1)   91.4 × 14.0

Fig. 16. Malvaceae.

    D3531C(8)   89.8 × 20.0

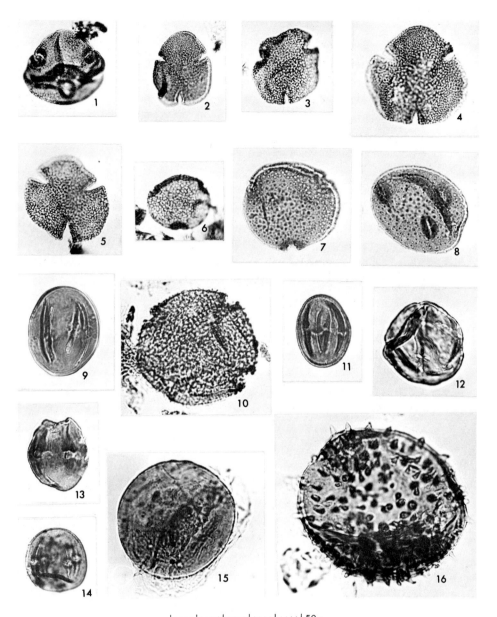

50 μ